EUL VERLAG

Reihe: Economy and Labour · Band 16

Herausgegeben von EUR ING Prof. Dr.-Ing. Hans-Georg Nollau
FBCS, Roßdorf

Prof. Dr.-Ing. Hans-Georg Nollau
Dipl.-Bw. (FH) Carmen Zech

Forecasting: A Challenge for True Statisticians

Bibliografische Information der Deutschen Nationalbibliothek

Die Deutsche Nationalbibliothek verzeichnet diese Publikation
in der Deutschen Nationalbibliografie; detaillierte bibliografische
Daten sind im Internet über <http://dnb.d-nb.de> abrufbar.

ISBN 978-3-8441-0009-9
1. Auflage Dezember 2010

© JOSEF EUL VERLAG GmbH, Lohmar – Köln, 2010
Alle Rechte vorbehalten

JOSEF EUL VERLAG GmbH
Brandsberg 6
53797 Lohmar
Tel.: 0 22 05 / 90 10 6-6
Fax: 0 22 05 / 90 10 6-88
http://www.eul-verlag.de
info@eul-verlag.de

**Bei der Herstellung unserer Bücher möchten wir die Umwelt schonen. Dieses
Buch ist daher auf säurefreiem, 100% chlorfrei gebleichtem, alterungsbestän-
digem Papier nach DIN 6738 gedruckt.**

Preface

At all times, looking into the future and knowing what is happening has been a dream of mankind. As a symbol for this attempt the Oracle of Delphi is the best proof and until today the Delphi-Method is an important decision support tool.

Despite all methods and procedures a high level of responsibility, seriousness and professionalism of all the involved people is an absolute necessity. Today, unfortunately we often have the situation very precisely described by NICHOLAS TALEB. He claims that those who are putting society at risk are "no true statisticians", merely people using statistics either without understanding or in a self-serving manner. This is not a joke, this is criminal!

In the present contribution, the 16th volume of the publication series "Economy and Labour" with the title "Forecasting: A Challenge for True Statisticians", a scientific, well proofed method of mathematical statistics for Time Series Analysis and Forecasting is presented. It is one of the mathematically oriented methods and procedures of Customer-oriented-Holistic-Netted-Logistics CHNL described in this publication series, such as

- ABC-Analysis, (Vol. 5);
- Controlling, (Vol. 8);
- Document-Management, (Vol. 12);
- Project-Management, (Vol. 10);
- Risk-Analysis and -Management, (Vol. 15);
- Simulation, (Vol. 13);
- Structured-Process-Reengineering, (Vol. 6);
- Total-Quality-Management, (Vol. 7);
- Vector-Benefitvalue-Analysis, (Vol. 14).

Like all these mathematical methods and procedures the forecasting tool is also described and its power is impressively shown by case studies using the SCA-Software system.

The present book is a revised version of the diploma thesis of Dipl.-Btrw. (FH) Carmen Zech at the University of Applied Sciences Regensburg Department BA, Logistics in 2002 controlled by Prof. Dr. Hans-Georg Nollau.

For the complete revision I thank Dipl.-Math. Walter Nollau.

Roßdorf, November 2010   Prof. Dr. H.-G. Nollau, FBCS

Table of Contents

Table of Figures

Table of Charts

Table of Abbreviations

ACF	Autocorrelation Function
AIC	AKAIKE I Criterion
AO	Additive Outlier
AR	Autoregressive
ARIMA	Autoregressive Integrated Moving Average
ARMA	Autoregressive Moving Average
BJM	BOX-JENKINS Method
CHNL	Customer Oriented Holistic Networked Logistics
CI	Confidence Interval
GUI	Graphical User Interface
IO	Innovational Outlier
IPO	Input-Process-Output
JIT	Just in Time
KGVL®	Kundenorientierte Ganzheitliche Vernetzte Logistik
LS	Level Shift
MA	Moving Average
PAC	Partial Autocorrelation
NSF	National Science Foundation
SARIMA	Seasonal ARIMA
SARMA	Seasonal ARMA
SCA	Scientific Computing Associates

TC   Temporary Change
TS   Time Series
TSA   Time Series Analysis

1 Introduction to the Customer Oriented Holistic Networked Logistics CHNL[1]

In the past the importance of logistics was constantly growing. The main reason is, that a solid performing logistics field gives a competitive edge in contest, which is of fundamental interest for a company. The permanent improvement and valorization of the products and production processes are the precondition to obtain advantage in this field.

Every advantage can be copied after a while and therefore the company has to become a moving target to get as fast new ideas as formers are copied. Single improvements are not enough for a constant competitive edge, they just give a short term of betterment in effects. Therefore it is essential to realize constant improvement if the company also wants to be a successful competitor in the future. Many improvements may be realized in the logistics chain. The costs will decrease effectively and great advantages in the market may be obtained.

For the logistics chain it is very important to know which demand has to be planned for the future. The problems of stock-keeping will thereby be minimized. The production won't need any overstocking, JIT can be realized in the holistic supply chain and not just set up as a solution on an island. Thereby a working recycling logistics can be created, without being overloaded by the amount of waste. Signals will be given early enough to realize borders of satisfaction within different product groups and therefore red numbers will be avoided.

There are already tools which make the demands of the future planable. The solution of the problems may be found in the analysis of time series and their forecast.

In the present publication the difficulties of time series analysis (TSA) will be discussed. The analysis will be made with the help of tools, which make it easy to create reliable forecasts.

But first we will give some basic information about the KGVL® (CHNL) and the historical development of the problem which is discussed in the present publication.

Detailed explanations about KGVL® (CHNL) and the practical meaning will be found in [N00].

---

1 CHNL is the English version of the German abbreviation KGVL® (Kundenorientierte Ganzheitliche Vernetzte Logistik) and means Customer Oriented Holistic Networked Logistics.

1.1   Contents of the CHNL

Modern logistics works with the whole information- and material flow. Before the material can flow, the information must flow in. The different fields of logistics consist of

- Provision
- Production
- Distribution
- Recycling

which are called the four pillars of logistics. The objects of logistics are not only materials in the common use. Moreover logistics must be explained in a more holistic manner. It is called the manner of the four M's

- Manhood
- Money
- Machines
- Material

The task of logistics is the solution of the problem to provide the companies with goods and services. This provision should be realized in the manner of the 7 Rs'.

The 7 Rs' of the logistics task are listed below. It means to keep ready

- the right amount
- of the right objects
- at the right place
- with the right information
- at the right time
- in the right quality
- to the right costs.

The customer has to be in the middle of every activity within the company. The task of the "Rs" has to be fulfilled to the satisfaction of the customer.

The principle of the logistics flow shows the dynamic in the logistics. The implementation of the principle will be explained as follows.

The elements are

- Market based steering
- Flowequitable structure organization
- Qualified staff

- Permeable information systems
- Processable structures
- Ruleable processes

Market based steering means a production which is triggered by the customer, but a 100% ability to supply has to be warranted. An overproduction is therefore impossible.

Flowequitable structure organization stands for flat structures. Response and competence have to be phrased in a clear way. Additionally the organization is product orientated.

Qualified staff doesn't just mean specialized knowledge rather it is flexible in its deployment, wants to take response and shows a lot of commitment.

On time delivered information is basic for permeable information systems. Therefore the company data have to be available on a networked system. If a permeable information system is realized, the effectiveness of the processes inside the company can speed up.

A flat structure of the single working steps, the latest possible change of the different models and human production facilities are the signs for a processable structure.

Ruleable processes are qualified with a secure and reconstructionable production.

Those three topics together, logistics-task, logistics-chain, principle of logistics flow combined represent the Client Orientated Holistic Networked Logistics CHNL (KGVL®)

1.2   Historic Development of the Problem

The problem discussed in the present publication originates from different earlier works. G. BAIER worked on the integration of forecast tools in standard software and mentioned their importance in [Ba00]. Suggestions for the problems of creating time series and data management in a data warehouse and for saving the results of forecasting time series out of the SCA-System in a data warehouse can also be found in this thesis.

J. HIRN has to deal with the problem of forecasting spare parts in [Hi00]. He suggested as forecast methods BOOTSTRAP and the CROSTON-Method. Finally NOLLAU and

HAUSER point out in [NoH01] the possibility of analyzing time series of demands with the help of ARIMA models.

Note: With regard to the fact, that the present publication is written in the English language, we want to mention that you will find standardized expressions of terms in the logistics field in the DINNORM 12777 in the languages German, English and French. This standard makes the international communication easier in the logistics field.

2   Basics of Time Series and their Forecast

Time series can be of regular or irregular kind. In the following chapters the forecast of regular time series will be discussed with the BOX-JENKINS method and the SCA software as a tool. In the Appendix 5 we will give a short introduction to methods of forecasting irregular time series. But at first we will give some common annotation for time series and forecasting procedures.

2.1   Fundamentals of Time Series

Theoretically time series can be separated into four basic components, which are associated in additive or multiplicative form:

Trend T	$\rightarrow$	basic tendency of the time series, long-term trend
Cycles C	$\rightarrow$	middle term periodical fluctuation
Seasonal fluctuation S	$\rightarrow$	steady periodical fluctuation
Rest component R	$\rightarrow$	random fluctuation

Therefore the time series can be explained as a function $y = f(T, C, S) + R$.

For the forecast of the time series it is important to figure out

- Which components exist in this species?
- Which attributes do the components have?
- How are they associated?

Annotation: Explanation of the terms season and period

In the literature about business and economics the term period in context with time series is often used as point of time, where the measurement happened, for example: turnover:

Turnover at the point of time 01/01/2000: $y1=$ turnover for period 1

Turnover at the point of time 02/01/2000: $y2=$ turnover for period 2

Also intervals of time are stated in amount of periods. For example: time from 01/01/2000 to 03/01/2000 (included) accords 3 periods.

This kind of using the term period is not in harmony with the mathematical meaning of period. The mathematical term period T is used in a periodical function f (t). T is the number with the following proberty:

$$f(t+T) = f(t) \qquad \forall \ t \text{ in the domain of the function,}$$

which means that, if the argument of the function f(t) gets enlarged (or reduced) with the period T, we get the same value of the function again. That means the process of the function repeats after the period T.

Neither we will break against the mathematical definition nor against the used term in the business administration. Therefore we will avoid the term period in the present publication. This is possible without any difficulties, because we can use the term season and seasonality, if periodicity in the mathematical sense is meant. We will follow this behaviour in this publication.

The amount of the time measurements, which are included in one season, is called length of season. This means, that the time series repeats itself (approximately) after those sections.

Example:

	Length of Season	**Duration**
Monthly measurements	12 months	1 year
Weekly measurements	52 weeks	1 year
Quarterly measurements	4 quarters	1 year
Daily measurements	7 days	1 week

Table 1: Lengths of seasons

2.2   Characteristics of Time Series

Time series can be characterized in two different ways:

Firstly in the length of the data series, which means the number n of measurements
Secondly in the appearance of the measurements contained in the data

2.2.1 Arrangement with the Length of the Time Series

	EFFECTIVE NUMBER OF MEASUREMENTS	
	NON SEASONAL $(p,d,q)-$ MODEL $N_{EFF} = N - d - p$	SEASONAL$(p,d,q)X(P,D,Q)-$ MODEL (LENGTH OF SEASON s) $N_{EFF} = N - d - p - s*D - s*P$
SHORT SERIES	$10 \leq n_{eff} < 50$	$2s \leq n_{eff} < 5s$
MIDDLE SERIES	$50 \leq n_{eff} < 100$	$5s \leq n_{eff} < 10s$
LONG SERIES	$n_{eff} \geq 100$	$n_{eff} \geq 10s$

Table 2: Characteristics of ARIMA/SARIMA Models

The different model types will be discussed later in this chapter in sections 2.4.2 and 2.5.

Note: ARIMA/SARIMA-Models are suitable for middle- and long series. For short series at the lower edge ($n_{eff} \approx 10$ respectively 2s) other methods should be used, for example Exponential Smoothing ([SPP98]).

2.2.2 Arrangement with the Kind of Appearance of the Measurements

As mentioned before there are irregular and regular time series. Regular time series have the property that nearly every measurement y_t is $\neq 0$, if anyways $y_t = 0$ then in very few points of time. A forecast method for regular time series is for example the BOX-JENKINS Method (BJM).

To the contrary irregular time series have the property that $y_t = 0$ appears very often and also repeatedly. An example for irregular time series is the demand of spare parts. Forecast methods for irregular time series are

- Method of CROSTON
- BOOTSTRAP-Method

2.3 Noteworthy Things with Forecasting Time Series

Basically a forecast is not an exact number, which will be reached in the future. So it is important that a good forecast is completed with an estimation of the error. Furthermore it is important to know how long the forecast horizon will be, the longer the forecast horizon is, the inexacter the forecast figure will be.

Additionally every kind of information, which is known, should be included in the forecast, so it is important that facts, which may influence the forecast, are taken care of, even if they are not mathematically.

Forecasts can be of intuitive kind or made with extrapolation, which means, that the time series is extrapolated into the future. In the present publication only the second kind of forecasting will be discussed.

The technique of forecasting should serve two demands. Firstly it should be stable against fluctuations, which are of stochastic kind, but secondly it has to react on influences, which are real changes.

The decision which forecasting method will be of best quality should be made by analysing the course of the data. A very helpful instrument is the graph of the time series. After recognizing the course of the data the choice of the forecasting method can be made.

Traditional methods for forecasting are Moving Averages, Exponential Smoothing and Regression, which won't be discussed in the present publication.

2.4   Theoretical Basics for Forecasting Time Series

2.4.1   Explanation of Autocorrelation and its Graphical Depicting

With the autocorrelation it is possible to examine a time series and its residuals. It is related to the correlation, which measures the coherences between two stochastic variables. The autocorrelation measures the coherence between a time series $\{ x_t \}$ and the time series $\{x_{t-k}\}$ which is moved for the time lag k. The empirical autocovariance for the time lag k is given by

$$\text{cov}(x_t, x_{t-k}) = 1/n \sum_{t=k}^{n} (x_t - \bar{x}) * (x_{t-k} - \bar{x})$$

Therefore the empirical autocorrelation is

$$r_k = \text{cov}(x_t, x_{t-k}) / \text{var}(x_t), \qquad -1 \leq r_k \leq 1$$

with the empirical variances of the time series x_t

$$\text{var} \, (x_t) = 1/n \sum_{t=0}^{n} (x_t - \bar{x})^2.$$

The autocorrelation coefficient can be interpreted as the PEARSON coefficient.

CORRELATION	STATISTICAL INTERPRETATION
$R_K = 0$	NO INFLUENCE
$R_K > 0$	POSITIVE INFLUENCE (BOOST)
$R_K < 0$	NEGATIVE INFLUENCE (DILUTE)

Table 3: Autocorrelation coeffizient

The graphic of the autocorrelation in dependence upon the time lag k (ACF = Autocorrelation Function) is called correlogramm. An example for the correlogramm is depicted below. This graph was made by the tool SCAGRAF that will be explained in chapter 4.

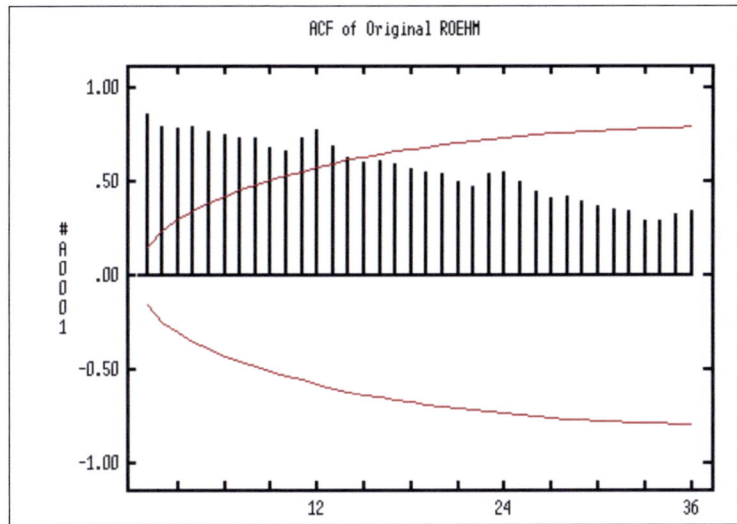

Fig. 1: Example of a Correlogramm

2.4.2   Explanation of Some Different Types of Models

Moving Average Model (MA-Model) ([G00])

If the measurement x_t is influenced by two random shocks at 2 points in time (t-1, t) then a model of the form

$$x_t = c + a_t - \theta_1 * a_{t-1} \qquad \text{for } |\theta_1| < 1$$

is valid. The influence of the shock a_t gets less, because of $|\theta_1| < 1$

Fig. 2: Influence of Shocks in a MA(1)-Model

If a shock effect of 3 points in time (t-2, t-1, t) is assumed, then follows

$$x_t = c + a_t - \theta_1 * a_{t-1} - \theta_2 * a_{t-2}$$

as a model.

Influence of shocks in a MA(2)-Model

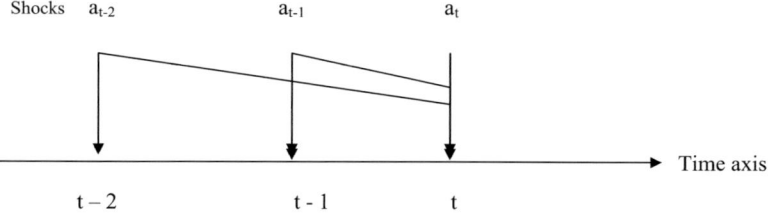

Fig. 3: MA(2)-Model

Every Moving Average Model has a specific course of the autocorrelation.

Therefore for a time series of a MA(1)-Model is valid

$$\text{cov}(x_t, x_{t-1}) = \theta_1 * \sigma^2_a$$

$$\sigma^2_a = \text{var}(a_t)$$

$$\text{var}(x_t) = (1 + \theta_1^2) * \sigma^2_a$$

Therefore the autocorrelation r_1 for the lag 1 is

$$r_1 = \theta_1 / (1 + \theta_1^2).$$

All other autocorrelations disappear.

Annotation: A cut of the autocorrelation function after the lag 1 indicates a time series model of the type MA(1). If the cut is after the second lag 2, the model seems to be of the type MA(2).

Autoregressive Model (AR- Model) ([G00])

If the observed value of time t - 1 immediately influences the value of the following point t, then this is called autoregression. In an autoregressive model of the order one AR(1), the actual value for the time t is equal to the value in the last point of time (t - 1) multiplied with the coefficient φ_1 plus a shock influence.

$$x_t = \varphi_1 * x_{t-1} + a_t.$$

An autoregressive model of order 2 also is influenced by the value in the point of time (t - 2), multiplied with coefficient φ_2

$$x_t = \varphi_1 * x_{t-1} + \varphi_2 * x_{t-2} + a_t.$$

The general AR(p)-model has the following form

$$x_t = c + \varphi_1 * x_{t-1} + \varphi_2 * x_{t-2} + \ldots + \varphi_p * x_{t-p} + a_t.$$

The covariance of lag 1 in an AR(1)-model is

$$\text{cov}(x_t, x_{t-1}) = \varphi_1 * \sigma^2_a$$
$$\text{var}(x_t) = \text{var}(a_t) = \sigma^2_a$$

Therefore the autocorrelation is

$$r_1 = \varphi_1.$$

Generalized, for an AR(p)-model is valid

$$r_k = \varphi_1 * r_{k-1} = \varphi^k_1.$$

For $|\varphi_1| < 1$, the series of the autocorrelations decreases, whereas a cut after a certain lag like in the MA model does not appear. Here the autocorrelation is not useful to get the maximum lag p. Therefore the series of partial autocorrelations (PAC) was developed. For an AR(1) model the PAC of order 2 will be much smaller than the PAC of order 1. You will find more details in ([Ch82]).

Mixed Model Types (ARMA/ARIMA SARIMA-Model)

If the time series is influenced by short-term shocks and also autoregression can be observed, mixed models of the type ARMA are assumed. An ARMA (1,1) model will be of the form

$$x_t = c + \varphi_1 * x_{t-1} + a_t - \theta_1 * a_{t-1}$$

The relation between an ARMA process and an ARIMA process is created with differencing and summation respectively.

ARIMA process X_t $\Rightarrow$ ARMA process $Y_t := X_t - X_{t-1} = (1-B)X_t$

Differencing

$X_t = X_0 + \sum^t_{k=1} Y_k$ $\Leftarrow$ Y_t

Summation

Differencing has to be made to get a stationary series.

A SARMA- or a SARIMA-model is a model including seasonality. For example a SARIMA $(p, d, q) \times (P, D, Q)$ model with $p = q = 1$, $d = 1$, $P = Q = 1$, $D = 1$ and seasonality $s = 12$ will be of the following form in operator writing

$$(1-B^{12})(1-B)x_t = [(1-\Theta_1 B^{12})*(1-\theta_1 B)/ (1-\Phi_1 B^{12})(1-\varphi_1 B)] * a_t$$

For more details see [LML01].

2.5   Outliers

Outliers may be explained as unexpected discontinuities that occur within a time series. These discontinuities are not compatible to the regular pattern within a time series and often introduce bias in the modeling and forecasting stage. Business and industrial data often include outliers. For example, a company's sales may be affected by a natural disaster or a strike. Those unexpected events make problems in forecasting. A statistical model is strictly based on the past history of a series. When models are used for forecasting (i. e. ARIMA, Exponential Smoothing etc.), it is assumed that the regular patterns exhibited in the past of a series will be continued in the future (i. e. through the forecasting time interval). With this in mind, it should be every forecasters objective to identify a model that performs well under normal data conditions without contamination by unexpected events. Outliers that occur in the forecasted time interval affect the statistics used to evaluate forecast performance. If outliers are not addressed, the model choice may be biased.

There are four different types of outliers considered in the SCA-System

Additive Outlier (AO)	Affects a series for one time interval only
Innovational Outlier (IO)	Affects some values after its occurrence
Level Shift (LS)	Affects a series at a given time, and the effect becomes permanent
Temporary change (TC)	Similar to IO, but the effect decreases.

3 Time Series Analysis of Regular Time Series with the SCA-System

To create forecasts of high quality, in the past it was necessary to have the knowledge of experts. Especially forecasting with ARIMA-models is of great extent.

With the SCA Software, especially the SCA-WORKBENCH ([Hu91], [SCA01], [SCA95], [LML01], [LML00], [L00]) we have a tool, which makes it easy to create high quality forecasts with Time Series Analysis (TSA), because it is made for users, which are just interested in forecast figures. The user just has to know some common statistical knowledge, which is fundamental to understand the forecast with time series. The use of the SCA-Software and the necessary statistical understanding will be explained in this chapter. The forecast tool of the SCA-System works with the BOX-JENKINS Method. Firstly we shall explain the idea of this method and the used model type in a brief summary and after this the SCA-System will be explained.

3.1 Basics of the BOX-JENKINS Method (BJM)

Forecasting with this method proved to be very good in making decisions in short- to medium time terms. The technique of this method is based on building a model with differences as filters to reach stationarity of the series. The BJM is based on three steps.

- Identification of the model
- Estimation of the parameters
- Diagnostic Checking

The identification of the model is done with basic statistical methods to answer the questions about transformation of the time series and the degree of the differences to make sure that the time series is stationary and to get hints of the degrees for the operator polynoms in the auto-regressive and moving-average part of the model.

The estimation of the parameters is done with the maximum-likelihood method.

In diagnostic checking the estimated model the primary interest is, that the residual series follows a white noise process. All those three steps will be discussed with the help of the SCA-System, after explaining the ARIMA/SARIMA model which is used in the BJM.

3.2   The Univariate ARMA/ARIMA and SARMA/SARIMA model

ARMA/ARIMA-model ([LML00])

The characteristic of an autoregressive-integrated moving average (ARIMA) model will be shown in this paragraph. The time series Z_t, $t = 0, 1, 2, \ldots, n$ should follow an ARMA – model (autoregressive moving average). An ARMA (p, q) – model has the explicit form

$$Z_t - \varphi_1 Z_{t-1} - \varphi_2 Z_{t-2} - \ldots - \varphi_p Z_{t-p} = C + a_t - \theta_1 a_{t-1} - \theta_2 a_{t-2} - \ldots - \theta_q a_{t-q}.$$

Where $\{a_t\}$ is a sequence of random errors, which means independently and identically distributed according to:

$$a_t \sim N(0; \sigma_a),$$

normal distribution with expectation 0 and standard deviation σ_a.

In order to derive the operator form of the ARMA (p, q)- model we need the back shift operator B, which is defined as follows

$$BZ_t = Z_{t-1}; \quad B^2 Z_t = B(BZ_t) = Z_{t-2};$$
$$B^k Z_t = Z_{t-k.}$$

Therefore

$$Z_t - \varphi_1 BZ_t - \varphi_2 B^2 Z_t - \ldots - \varphi_p B^p Z_t = C + a_t - \theta_1 Ba_t - \theta_2 B^2 a_t - \ldots - \theta_q B^q a_t.$$

This can be abbreviated to

$$\varphi(B) Z_t = C + \theta(B) a_t$$

with

$$\varphi(B) = (1 - \varphi_1 B - \varphi_2 B^2 - \ldots - \varphi_p B^p), \qquad \text{autoregressive operator (-polynomial)}$$
$$\theta(B) = (1 - \theta_1 B - \theta_2 B^2 - \ldots - \theta_q B^q), \qquad \text{moving average operator (-polynomial).}$$

This is the operator form of the ARMA(p, q) model with p as the degree of the autoregressive operator and q as the degree of the moving average operator. This model also can be expressed as

$$Z_t = \mu + [\theta(B) / \varphi(B)] * a_t \qquad \text{with } \mu = C/(1 - \varphi_1 - \varphi_2 - ... - \varphi_p).$$

If the series is not stationary (i. e. has no fixed mean level and constant standard deviation), then the autoregressive portion of the ARMA(p, q) model must include a stationarity inducing operator. This is accomplished by means of a differencing operator, with the form (1-B). So instead of modeling the non-stationary series Z_t, the series

$$(1-B) Z_t = Z_t - Z_{t-1}, \qquad t = 1, 2, ..., n$$

will be created.

The model, which will be considered, will be an autoregressive-integrated moving average model ARIMA(p, d, q), with d giving the order of the differencing operator. This model is of the operator form

$$\varphi(B)(1-B)^d Z_t = C + \theta(B) a_t \qquad \text{or}$$

$$(1-B)^d Z_t = \mu + [\theta(B) / \varphi(B)] * a_t \qquad \text{with} \quad \mu = C/(1 - \varphi_1 - \varphi_2 - ... - \varphi_p).$$

SARMA/ SARIMA model   ([L00])

The form of a multiplicative seasonal ARIMA model can be expressed as

$$\varphi(B)\Phi(B^s)(1-B)^d(1-B^s)^D Z_t = C + \theta(B)\Theta(B^s)a_t$$

or

$$(1-B)^d(1-B^s)^D Z_t = \mu + [(\theta(B)\Theta(B^s)) / \varphi(B)\Phi(B^s)] * a_t$$

where

$$\Phi(B^s) = 1 - \Phi_1 B^s - \Phi_2 B^{2s} - ... - \Phi_P B^{Ps}, \text{ and}$$

$$\Theta(B^s) = 1 - \Theta_1 B^s - \Theta_2 B^{2s} - ... - \Theta_Q B^{Qs}$$

$$\mu = C/[(1 - \varphi_1 - \varphi_2 - ... - \varphi_p) * (1 - \Phi_1 - \Phi_2 - ... - \Phi_p)]$$

For example the time series ROEHM (see 3.5) would be of the following form.

SARIMA (0, 1, 1) x (0, 1, 1) with

$(1\text{-}B)(1\text{-}B^{12}) *Z_t = C + (1\text{-}\theta B) * (1\text{-}\Theta B^{12}) * a_t.$

Note: The values of D, P and Q are mostly 0 or 1. For more details see Appendix 1.

3.3   Basics of the SCA-Software

The SCA-Software is a holistic instrument to solve statistical problems. Especially in logistics we have different problems, which have to be solved with mathematical instruments. The use of the SCA-system will just be explained in the field of TSA using the BJM. Time series can be found in each of the four logistics pillars according to paragraph 1.1. In chapter 3 and 4 we will show the use of TSA with one example taken from the field of distribution logistics (time series ROEHM, section 3.5) and two examples from the field of the waste logistics (time series SAM1, section 6.1 and time series SAM2, section 6.2).

The SCA-Software package also includes instruments for

- Descriptive statistics and correlation
- Displaying of location, dispersion and distribution
- Cross tabulation
- Comparing two samples
- Analysis of variance
- Linear regression analysis
- **BOX-JENKINS time series modeling and forecasting** *(which is used in the present publication)*
- Nonparametric statistics
- Distribution and model simulation
- Analytic functions and matrix operations
- A unique graphic package with many possibilities for graphics depicting.

The development of the SCA product began in 1981. It is designed by LON-MU LIU with the assistance of the SCA programming staff ([LML00]). The development of the 16-bit version of the SCA-System was of fundamental interest to use statistical software on a personal computer, which is able to create high quality forecasts. The next generation for PC use was the

32 bit version with new capabilities. But in both versions it was necessary to have very good statistical knowledge to be able to use this software in a sensible way. Now with the development of the SCA-WORKBENCH a system was created with an easy to use surface, also for users who just have basic knowledge in the statistical area of forecasting. It provides a comprehensive feature to benchmark time series models and evaluate forecasting performance.

3.4   Installation of the SCA – Statistical System

PC-Installation of the SCA-System

The requirements for the SCA-WORKBENCH system are

- Pentium PC or higher
- CD-ROM drive and 3.5'' floppy drive
- Minimum 10MB of hard disk space for installed files
- Microsoft Windows 95, 98, NT, 2000 or higher
- Minimum 32MB of RAM (64MB or more recommended)

The installation program is completely automatic and self-guided. The setup program is located on the SCA CD-ROM. It is possible to select the components of the SCA Software, which should be installed. It also gives the choice of where to install the SCA products. During the installation the setup program will demand to insert the SCA disk#1. This seems a kind of weird, but makes sense after knowing what happens. On the disk is a check count program, which tells the setup program how often somebody is allowed to install the program and how much setups are left. It also sets the counter back, if the SCA software is uninstalled. Two setups are possible. The CD-ROM also includes an uninstall program, so it is possible to uninstall the software and install it on another PC. Therefore the check count program on the disk is needed again.

There are also different documentations provided on the CD-ROM in PDF-format. They are updated to the current version of the SCA Software. Those documents are

- Supplemental Documentation for the PC SCA Statistical System
- SCA GRAPHIC PACKAGE User's Guide
- SCA WORKBENCH User's Guide
- Collection of Time Series DATA (WP109)

Finally it is possible to create a shortcut on the desktop. An icon can be created by right-clicking anywhere at the desktop, then choosing from the appearing menu the option *New* and then selecting *Shortcut* in the pop up menu. Windows will open a dialog box and prompt to enter a command line. There SCAW32 has to be entered in the command line with the appropriate parameter specification, where the SCA Software was saved to, during the installation (for example: C:\SCAW32). The created shortcut will look like the following icon

SCA WorkBench 2000.lnk

Fig. 4: Icon for SCA-WORKBENCH

Network Installation of the SCA-System

The first step is to insert the SCA CD-ROM and execute the setup program. Within this setup program, you will be prompted to indicate whether you are installing the PC SCA System on a local hard drive or a network drive. It is necessary to select network installation for using the SCA software also from other computers. Otherwise it will deny the access to other users.

The files have to be made under the SCA installation directory and subdirectory shareable. The SCA installation directory also has to be included in the path of the user's LOGIN script file.

3.5   Working with the SCA-System According to Input-Process-Output (IPO)

The present publication will explain the SCA-System in the order of the IPO principle. Since the SCA-software offers the possibility to work with three kinds of process tools, the IPO-principle is explained for every environment itself. The three different processes can be explained as follows

System command driven by paragraphs   RUN SCA 32 BIT SYSTEM INTERACTIVLY
System command driven by macros   RUN SCA 32 BIT SYSTEM WITH MACRO
Input window system (automatically)   WB, APPS   AUTOMATIC MODELING
                                         WB, APPS   USER DEFINED MODELING
(WB = Workbench, APPS = Menu APPLICATIONS)

Please find a chart below where you can find the different steps of the IPO principle for each environment.

	Input	**Process**	**Output**
System command driven by paragraphs	page 26f	page 27-40	page 28
System command driven by macros	page 42, 54-59	page 60-66	App. 2
Input window system (automatically)	page 42, 54-59	page 42-46,50f	page 46-49

Table 4: Overview of the IPO Principle

We will work with three different time series taken form the logistics field. With the time series ROEHM we will explain the different functions and options of the SCA System. The other two time series are SAM1 and SAM2. We will use them to show the whole forecast process with two examples. But first we will give some information about the different data-sets.

Name of TS	Meaning	Aggregation	Number of measure-ments
ROEHM	Turnover data	Monthly	155 (long-series)
SAM1	Amount of waste	Monthly	113 (middle-series)
SAM2	Additional quantity to amount of waste	Monthly	65 (short-series)

Table 5: Data-Cases

Because the time series ROEHM is used to explain how the SCA System works, we will depict the whole series now. SAM1 and SAM2 are depicted in chapter 6, when they are used as examples to create forecasts.

40930 42157 45118 42477 46713 37055 44186 42259 48506 44404 49475 54469 44505 50010 49674 51855 50061 42521
47337 51527 65477 57144 59917 53437 1097 52473 58098 60569 56631 40642 52584 52381 72881 67493 67044 58751
57951 53819 47682 53722 62563 53950 45169 50142 64842 64415 62627 65591 59014 60736 52459 67380 71128 63357
50581 50733 62097 75513 69218 65436 69893 66436 58176 62963 61890 61543 52331 60538 5961373991 72627 69765
73305 69039 66818 72441 73264 74370 60276 70232 74129 81001 70159 96764 74636 83122 80009 73393 88407 88584
62441 72983 81014 97336 88744 90728 84642 93958 79324 83911 98252 89841 64508 78214 96092 87518 105516 85963
95749 91359 74336 87154 99349 83082 68199 82508 91902 41986 45791 53907 43556 46953 47963 43696 41007 46848
43791 45619 34774 42258 45003 52523 48653 46137 50070 98661 96598 88627 96332 97254 79244 97116 99335 91661
70662 86012 96677 109153 89718 99822 103851 95995 99357 98639 94908

Number of measurements n = 155

Fig. 5: Time Series ROEHM

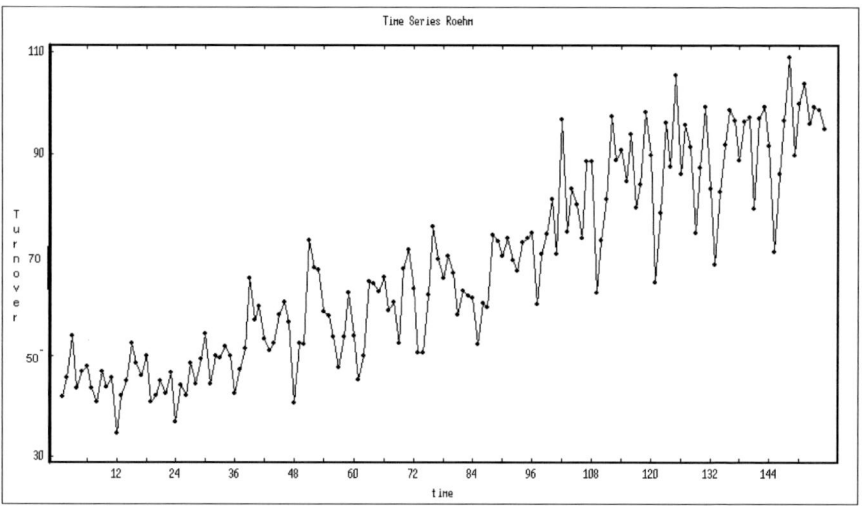

Fig. 6: Plot of Time Series ROEHM

Note: This plot was created with SCAGRAF, see chap. 4. The ordinates must be multiplied by 1000.

Whereas there are three different kinds of processing the data, the SCA-System itself has just two different working surfaces. We first will explain the different surfaces. After the surfaces are explained we will show how the data process works in the three possibilities with the time series ROEHM and compare the results of each tool. In the end we will give a recommendation in chapter 6 how to use the SCA-System and which possible forms may be used to show the results of the analyzed time series.

3.5.1   The System Command Driven by Paragraphs

The 32-bit version of SCA Expert is the environment which is driven by commands which in terms of the SAC-systems are called *paragraphs*. It reacts to commands (paragraphs) given by the user. All instructions must be ended by a carriage return. Important to know is, that all command lines of the SCA-System are preceded by the symbol '-- ', which is the SCA System prompt.

To access the system on a mainframe computer, the command **SCA** has to be entered. On the PC you just have to double click the created icon or call it from the menu option in MS-WINDOWS, like below.

START – PROGRAMS – 32bit SCA Statistical System

For entering the 32-bit version from the SCA-Workbench, the workbench has to be started by double clicking the icon of SCA, which was created. The main menu bar will be displayed on the screen and look like the bar below.

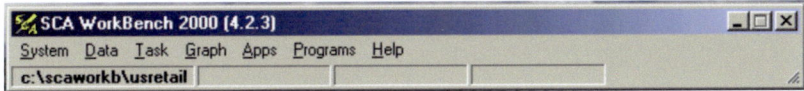

Fig. 7: Menu bar of SCA-WORKBENCH.

Menu	Item Description
System	The System menu contains preference settings for SCA WORK-BENCH. It also provides sub-menu items to run the command driven SCA-System interactively or with macros.
Data	The Data menu item contains tools to open a spreadsheet file for example in Microsoft Excel spreadsheet applications. It also contains a utility to generate a SCA Data Macro from the data selected in a spreadsheet.

Task	This item provides capabilities to edit SCA macro procedures, create and edit Macro-only Tasks and Spreadsheet Tasks, and assemble tasks in projects. This item will be explained in more detail later (see section 3.5.3.1).
Graph	The Graph menu item provides access to the SCAGRAF program.
Apps	The Apps menu item contains capabilities for building time series models using an enhanced graphical user interface as well as capabilities to perform model benchmarking using rolling forecasting. This item will be explained in more detail later (see section 3.5.2).
Programs	Here applet programs can be added/removed, which may be executed from SCA-Workbench.
Help	The Help menu item contains a help program for the workbench and also for commands.

For calling the command driven System 32-bit Version the menu *System* has to be selected.

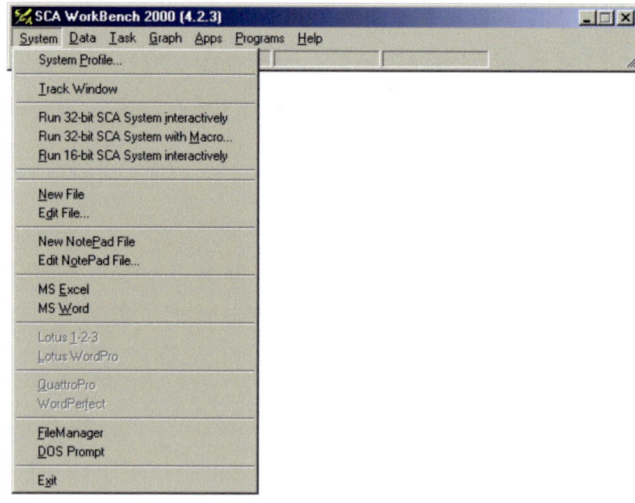

Fig. 8: Surface of SCA-WORKBENCH and the Options of the Menu "System"

The menu "System" contains the submenus

Run 32-bit SCA System interactively: Launches the 32-bit SCA System in interactive mode.

Run 32-bit SCA System with Macro: Executes a macro-procedure in the 32-bit SCA System
and provides a convenient environment to modify macro
files, to submit macros to the SCA System and to review
output results.

After choosing the 32-bit version *interactively* some short descriptive information appears. This set of information includes SCA release version, product names and so on (see Fig. 9). This information is followed by a SCA prompt ("--"). This is an indication, that SCA commands now can be entered. The surface is similar to the DOS surface, where the commands also have to be entered directly into the screen.

Commands are also called statements in the SCA System. It distinguishes between two groups of statements, namely

analytical statements   used for operations (for example LNY=LN(Y))
english-like statements   used to accomplish most operations (for example ACF,
IARIMA, ESTIM).

To exit from a SCA session the command STOP has to be entered.
To quit an actual command, the command QUIT must be entered.

The surface of the SCA 32-bit version of SCA EXPERT, where the commands have to be entered, is depicted below.

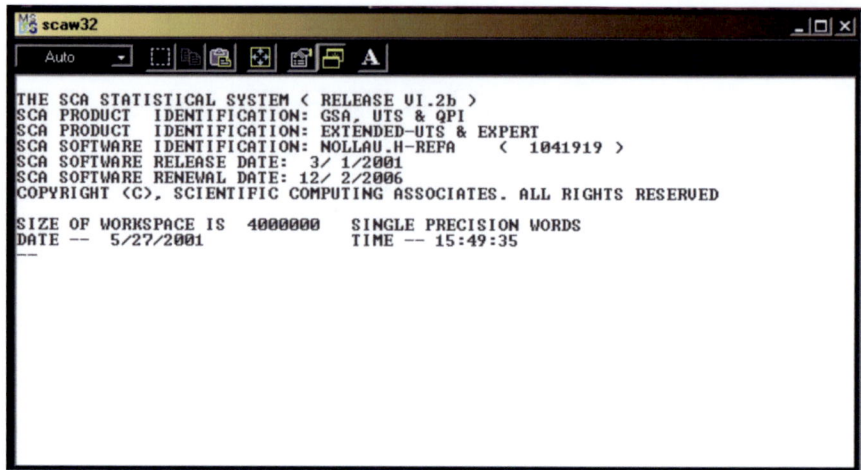

Fig. 9: Surface of the SCA 32-bit System

Entering Data Into the System From the Terminal

The data can be entered directly from the terminal. The data can be entered **case-by-case**, which means, the first case for all the variables, then the second case for all the variables and so on, is entered.

Example for Entering Data Case by Case:

                INPUT PRICE, MONTH

The system will prompt for the data. Every case has to be entered in a new line (record), for example:

73 10

78 11

85 12

90 13

91 14

87 15

86 16

91 17

75 18

65 19

END OF DATA

The systems message will be

> PRICE, A 10 BY 1 VARIABLE, IS STORED IN THE WORKSPACE
>
> MONTH, A 10 BY 1 VARIABLE, IS STORED IN THE WORKSPACE

Entering Data From a File

If data from an external file should be entered into the workspace, it has to be a "flat file". Flat file means ASCII-data in form of a table which is built of rows and columns. This data may be created in a text editor. Most time flat files contain only one data set, or one set of case-by-case data records. The command for entering a file into the system is

> INPUT variable name. FILE IS "file-name"

For example: INPUT ROEHM. FILE IS "A:ROEHM.DAT"

The command END OF DATA is not necessary, but should nevertheless be given.

Model Identification

Aim of this topic is to determine appropriate values for p, d and q for the ARIMA (p, d, q) model. It is not necessary to find the unique model for the time series, but it should be possible to restrict (p, d, q) to a limited number of models. Significance will be determined during the estimation and diagnostic checking stages.

The starting point of the model identification should always be the graph of the time series. The graph of the time series ROEHM is depicted in Fig. 6. The examiner will recognize the increase of the spread in the series. This is a hint to transform the data with a ln-transformation.

Note: The ln-transformation was not done in this case, because the result of the estimated model gave very good forecasts with small standard errors.

The linear trend in the series is an advice to difference the data (with $d = 1$ (operator $(1-B)$)) to get a stationary series. Because it seems that there is also a seasonality included it may be useful to difference the time series also with a seasonal difference ($D = 1$, operator $= (1-B^{12})$).

If the time series is not stationary, it has to be differenced, as mentioned above. If a time series is non-stationary, its ACF (autocorrelation function) will be high for a number of lags. For computing and displaying the sample ACF, the command

ACF name of time series. MAXLAG IS number

has to be entered. ACF will give the autocorrelation as depicted below. MAXLAG will limit the computed lags.

The following commands were entered:

INPUT ROEHM. FILE IS 'A:ROEHM.DAT'
(Series ROEHM is loaded into the workspace)

ACF ROEHM. MAXLAG is 36

The output of the ACF- command on the screen in the SCA window is not complete, because it is truncated. The user will get the information of the last 15 rows. In order to show the whole information, the command REVIEW must be employed. This command opens a text editor including the information of the commands in the current SCA session. This information is stored in the file SCAOUT.OTP. Below is an example of the ACF- paragraph and its depicting with the REVIEW command in the editor.

To open the editor and the SCAOUT.OTP file, the command

REVIEW

has to be entered, without any other information.

Output of ACF ROEHM with REVIEW command is as follows:

NAME OF THE SERIES . . . .   ROEHM
TIME PERIOD ANALYZED . . . . 1 TO   155
MEAN OF THE (DIFFERENCED) SERIES . . . 67396.6094
STANDARD DEVIATION OF THE SERIES . . . 18816.6016
T-VALUE OF MEAN (AGAINST ZERO) 44.5926
AUTOCORRELATIONS

```
1- 12    .86 .79 .78 .79 .76 .75 .73 .72 .68 .66 .73 .77
ST.E.    .08 .13 .16 .18 .20 .22 .23 .25 .26 .27 .28 .29
Q        116 216 312 413 508 599 687 774 850 923 1012 1113


13- 24   .68 .62 .60 .60 .59 .56 .54 .54 .49 .47 .54 .55
ST.E.    .31 .32 .32 .33 .34 .34 .35 .36 .36 .37 .37 .37
Q        1192 1258 1321 1385 1446 1501 1554 1606 1650 1690 1743 1799


25- 36   .49 .44 .41 .41 .39 .36 .35 .34 .28 .29 .32 .34
ST.E.    .38 .38 .39 .39 .39 .39 .40 .40 .40 .40 .40 .41
Q        1844 1881 1913 1945 1974 2000 2024 2047 2063 2080 2101 2125


      -1.0 -0.8 -0.6 -0.4 -0.2  0.0  0.2  0.4  0.6  0.8  1.0
           +----+----+----+----+----+----+----+----+----+
                         I
 1  0.86            +   IXXX+XXXXXXXXXXXXXXXXX
 2  0.79             +  IXXXXX+XXXXXXXXXXXXXX
 3  0.78            +   IXXXXXXX+XXXXXXXXXXX
 4  0.79            +   IXXXXXXXX+XXXXXXXXXXX
 5  0.76           +    IXXXXXXXXX+XXXXXXXXX
 6  0.75          +     IXXXXXXXXXX+XXXXXXXX
 7  0.73          +     IXXXXXXXXXX+XXXXXXX
 8  0.72          +     IXXXXXXXXXXX+XXXXXX
 9  0.68         +      IXXXXXXXXXXXX+XXXX
10  0.66         +      IXXXXXXXXXXXX+XXX
11  0.73         +      IXXXXXXXXXXXXX+XXXX
12  0.77         +      IXXXXXXXXXXXXX+XXXXX
13  0.68        +       IXXXXXXXXXXXXXX+XX
14  0.62        +       IXXXXXXXXXXXXXXX
15  0.60        +       IXXXXXXXXXXXXXXX+
16  0.60        +       IXXXXXXXXXXXXXXX+
17  0.59       +        IXXXXXXXXXXXXXX +
18  0.56       +        IXXXXXXXXXXXXX +
19  0.54       +        IXXXXXXXXXXXXX +
20  0.54       +        IXXXXXXXXXXXX  +
```

21	0.49	+	IXXXXXXXXXXXX	+
22	0.47	+	IXXXXXXXXXXXX	+
23	0.54	+	IXXXXXXXXXXXXXX	+
24	0.55	+	IXXXXXXXXXXXXXX	+
25	0.49	+	IXXXXXXXXXXXX	+
26	0.44	+	IXXXXXXXXXXX	+
27	0.41	+	IXXXXXXXXXX	+
28	0.41	+	IXXXXXXXXXX	+
29	0.39	+	IXXXXXXXXXX	+
30	0.36	+	IXXXXXXXXX	+
31	0.35	+	IXXXXXXXXX	+
32	0.34	+	IXXXXXXXXX	+
33	0.28	+	IXXXXXXX	+
34	0.29	+	IXXXXXXX	+
35	0.32	+	IXXXXXXXX	+
36	0.34	+	IXXXXXXXXX	+

Fig. 10: ACF of the Time Series ROEHM

The '+' sign depicts the limit of significance and the 'x' are the values of the autocorrelation. An autocorrelation is significant if it is greater than twice of its standard deviation. All values of the autocorrelations which cross the limit of significance are significantly different from zero and therefore they are large values. Values below the significance limit are small and therefore not significantly different from zero. In this example there are eight large values in the ACF. All values are positive and decrease slowly. It seems that this time series has to be differenced at least once. So lets put d = 1. Therefore the differencing operator (1-B) should be included in the model of this time series.

Model Specification and Estimation

Because of our look at the graph (see Fig. 6 and the previous section *Model Identification*) we expect a SARIMA (p, d, q) x (P, D, Q) model. For example we assume for the time series ROEHM a model of the form SARIMA(0,1,1) x (0,1,1). This model will be of the following form

$$(1\text{-}B)\,(1\text{-}B^{12})\,*Z_t = C + [(1\text{-}\theta B) * (1\text{-}\Theta\,B^{12})] * a_t.$$

Entering the command

TSMODEL NAME IS MODELROE. MODEL IS @

ROEHM(1,12)=(1-THETA1*B)(1-THETA12*B**12)NOISE

can specify this model with C = 0.

NOTE: Variable names in the SCA- System may not have more than eight characters.

For detailed information how a specified model is translated into the SCA-conventions see Appendix 1.

The name of the model is an individual name, which is saved in the workspace during the SCA session. It has to be different from the other variable names.

To estimate the model the command

ESTIM MODELROE. METHOD IS EXACT.HOLD RESIDUALS(resida).

has to be entered. If there is a MA-part in the time series, it is recommended to use the paragraph METHOD IS EXACT. The HOLD parameter takes care that residuals are maintained in the workspace for diagnostic checking. The following information about the model will be given.

```
THE FOLLOWING ANALYSIS IS BASED ON TIME SPAN   1   THRU   155

  NONLINEAR ESTIMATION TERMINATED DUE TO:
  RELATIVE CHANGE IN THE STANDARD ERROR LESS THAN 0.1000D-02

SUMMARY FOR UNIVARIATE TIME SERIES MODEL -- MODELROE
-------------------------------------------------------------------
  VARIABLE    TYPE OF     ORIGINAL     DIFFERENCING
              VARIABLE    OR CENTERED
                                         1      12
  ROEHM       RANDOM      ORIGINAL     (1-B ) (1-B )
-------------------------------------------------------------------

  PARAMETER   VARIABLE  NUM./  FACTOR  ORDER   CONS-     VALUE    STD     T
    LABEL       NAME    DENOM.                 TRAINT            ERROR   VALUE

    1 THETA1    ROEHM    MA      1       1     NONE      .8544   .0432  19.80
    2 THETA12   ROEHM    MA      2      12     NONE      .5514   .0722   7.63

  EFFECTIVE NUMBER OF OBSERVATIONS . .          142
  R-SQUARE . . . . . . . . . . . . .          0.894
  RESIDUAL STANDARD ERROR. . . . . .    0.612863E+04
```

Fig. 11: Output Results of the ESTIM Command in SCA EXPERT

Note: If the T-value for a certain parameter is below $|T| = 1.65$, than this parameter is not significant and the command TSMODEL has to be adjusted in the way to leave this parameter out.

Error Probability	Safety Probability	Criterion	Significance
$\alpha > 10\%$	$S \leq 90\%$	$|T| \leq 1.645$	not significant (-)
$5\% < \alpha \leq 10\%$	$90\% < S \leq 95\%$	$1.645 < |T| \leq 1.960$	weakly significant (*)
$1\% < \alpha \leq 5\%$	$95\% < S \leq 99\%$	$1.960 < |T| \leq 2.576$	significant (**)
$0.1\% < \alpha \leq 1\%$	$99\% < S \leq 99.9\%$	$2.576 < |T| \leq 3.291$	highly significant (***)
$\alpha \leq 0.01\%$	$S > 99.9\%$	$|T| > 3.291$	most highly significant (****)

Table 6: T-Values for Significance Tests

The estimated values of the different parameters can be taken directly out of the generated output data. The information for the parameters comes with the order, i.e. the number which describes the position of the parameter in the AR- or MA-operator polynomial respectively, the standard errors and T-values of the parameters. For example θ_1 has the value 0.8544 with the standard error of 0.0432 and a T-value of 19.80, which means the parameter θ_1 is most highly significant.

Diagnostic Checks of the Model

Two things have to be checked in this last step: firstly, whether the model does fulfill the requirements and secondly whether it does make sense. The basic assumption for ARIMA models is that the errors are independent and normally distributed, that means, the residuals follow a *white noise* process. If this is not true the model needs to be modified. The residuals can be checked in different ways. One possibility is to plot the residuals. They all have to be small (not significant). Another diagnostic check is the ACF of the residual series. If it follows a white noise process, then no autocorrelation should be significant. By entering the command

ACF RESIDA. MAXLAG IS 36

the following result is generated

```
NAME OF THE SERIES . . . . . . . . . .        RESIDA
TIME PERIOD ANALYZED . . . . . . . . . 14   TO   155
```

```
MEAN OF THE (DIFFERENCED) SERIES . . .      232.1988
STANDARD DEVIATION OF THE SERIES . . .     6109.1274
T-VALUE OF MEAN (AGAINST ZERO) . . . .        0.4529
```

AUTOCORRELATIONS

```
 1- 12    -.13  .01  .03 -.03  .05  .16 -.06  .11  .02 -.15  .06  .08
 ST.E.     .08  .09  .09  .09  .09  .09  .09  .09  .09  .09  .09  .09
 Q         2.5  2.5  2.6  2.7  3.1  6.8  7.5  9.3  9.3 12.6 13.1 14.1
13- 24    -.09  .07 -.09 -.10  .07 -.07 -.06  .08 -.00 -.23  .19 -.26
 ST.E.     .09  .09  .09  .09  .09  .09  .09  .09  .10  .10  .10  .10
 Q        15.3 16.2 17.5 19.1 19.9 20.8 21.5 22.6 22.6 31.6 37.6 48.9
25- 36     .01  .13 -.08  .05 -.04 -.13  .03  .03 -.15  .05 -.00 -.06
 ST.E.     .11  .11  .11  .11  .11  .11  .11  .11  .11  .11  .11  .11
 Q        48.9 51.8 53.0 53.4 53.7 56.6 56.8 56.9 60.9 61.4 61.4 62.1
          -1.0 -0.8 -0.6 -0.4 -0.2  0.0  0.2  0.4  0.6  0.8  1.0
           +----+----+----+----+----+----+----+----+----+----+
                                    I
 1  -0.13                      +XXXI   +
 2   0.01                      +   I   +
 3   0.03                      +   IX  +
 4  -0.03                      +   XI  +
 5   0.05                      +   IX  +
 6   0.16                      +   IXXXX
 7  -0.06                      + XXI   +
 8   0.11                      +   IXXX+
 9   0.02                      +   I   +
10  -0.15                      XXXXI   +
11   0.06                      +   IX  +
12   0.08                      +   IXX +
13  -0.09                      + XXI   +
14   0.07                      +    IXX  +
15  -0.09                      +  XXI    +
16  -0.10                      +  XXI    +
17   0.07                      +    IXX  +
18  -0.07                      +  XXI    +
19  -0.06                      +  XXI    +
20   0.08                      +    IXX  +
21   0.00                      +    I    +
22  -0.23                      X+XXXXI   +
23   0.19                      +    IXXXXX
```

```
24   -0.26                    X+XXXXI      +
25    0.01                    +    I       +
26    0.13                    +    IXXX +
27   -0.08                    +  XXI       +
28    0.05                    +   IX       +
29   -0.04                    +   XI       +
30   -0.13                    + XXXI       +
31    0.03                    +   IX       +
32    0.03                    +   IX       +
33   -0.15                    +XXXXI       +
34    0.05                    +   IX       +
35    0.00                    +    I       +
36   -0.06                    +  XXI       +
```

Fig. 12: ACF of the Residuals for the Time Series ROEHM in SCA-EXPERT

Two significant autocorrelations at lags 22 and 24 can be found in this figure and 3 others with values on the significance limit. So it seems that the model has to be improved, but we will not do that here and take the model as it is.

Forecasting an Estimated Model

Once we have accepted the model, we can forecast the series with the command

                    FORECAST name of the model. NOFS ARE number

NOFS will set the number of forecasts. The command

                    FORECAST MODELROEHM. NOFS ARE 4

gives the result

```
    ---------------------------------
     4 FORECASTS, BEGINNING AT   155
    ---------------------------------
     TIME     FORECAST   STD. ERROR   ACTUAL IF KNOWN
      156   95835.6503   6128.6314
      157   76370.2009   6193.2558
      158   90104.8560   6257.2128
      159  100768.0749   6320.5227
```
Fig. 13: Forecast of Time Series ROEHM

Four forecasts are given together with the standard error of each forecast.

Outlier Detection

To check if outliers are included in the data of the time series, to estimate the model parameters and to make forecasts when outliers are present, the following commands have to be used in the forecast session.

The command OUTLIER will detect the outliers in the time series, together with their types. The syntax for this command is

OUTLIER model name. TYPES ARE AO,IO,LS,TC

The following result was given for the model MODELROE

```
OUTLIER modelroe. TYPES ARE AO,IO,LS,TC

 INITIAL RESIDUAL STANDARD ERROR =  0.23311E+13
 ADJUSTED RESIDUAL STANDARD ERROR =  0.16440E+13

INITIAL RESIDUAL STANDARD ERROR =   6027.8
   TIME          ESTIMATE    T-VALUE    TYPE
   102           55166.28     3.49      IO
ADJUSTED RESIDUAL STANDARD ERROR =   5762.6
```
Fig. 14: Result of the OUTLIER Command for Time Series ROEHM

Alternative to the ESTIM paragraph the OESTIM paragraph may be used. The SCA System will simultaneously detect outliers and estimate their effect.

The syntax will be

OESTIM model name. METHOD IS EXACT.

The parameter METHOD IS EXACT may be useful, if there is a MA- part in the model.

The following result was given after entering the paragraph

OESTIM modelroe. METHOD IS EXACT

```
THE FOLLOWING ANALYSIS IS BASED ON TIME SPAN   1  THRU  155

SUMMARY FOR UNIVARIATE TIME SERIES MODEL -- MODELROE

-----------------------------------------------------------------------
VARIABLE    TYPE OF     ORIGINAL     DIFFERENCING
            VARIABLE    OR CENTERED
                                          1      12
 ROEHM      RANDOM      ORIGINAL      (1-B  ) (1-B  )
-----------------------------------------------------------------------

 PARAMETER   VARIABLE   NUM./  FACTOR  ORDER   CONS-      VALUE     STD     T
  LABEL        NAME     DENOM.                 TRAINT             ERROR   VALUE

  1 THETA1    ROEHM      MA      1      1      NONE       .8283    .0488   16.97
  2 THETA12   ROEHM      MA      2      12     NONE       .5125    .0773    6.63


SUMMARY OF OUTLIER DETECTION AND ADJUSTMENT

---------------------------------------
 TIME    ESTIMATE   T-VALUE    TYPE
---------------------------------------
  102  18313.927     3.72      AO
  125  16790.133     3.39      AO
---------------------------------------

TOTAL NUMBER OF OBSERVATIONS. . . . . . . . . . . . .        155
EFFECTIVE NUMBER OF OBSERVATIONS. . . . . . . . . . .        142
RESIDUAL STANDARD ERROR (WITHOUT OUTLIER ADJUSTMENT). .  0.645085E+04
RESIDUAL STANDARD ERROR (WITH OUTLIER ADJUSTMENT) . . .  0.592965E+04
```

Fig. 15: Result of the OESTIM Command

Note: Important to know about this command is, that also a number of variables can be created. which are useful for further analyses and the depicting of outliers. The complete list of available parameters for this command is

```
OESTIM  MODEL model-name.                          @
        TYPES ARE w1, w2, - - - .                  @
        DELTA IS r.                                @
        OSTOP ARE MXOUTLIERS(i1), CRITICAL(r),     @
              MXESTIM(i2).                          @
        NEW-SERIES IN v1, v2, v3, v4, v5.          @
        METHOD IS w.                               @
        STOP ARE MAXIT(i), LIKELIHOOD(r1),         @
               ESTIMATE(r2), STDEV(r3).            @
        OADJUSTMENT IS w.                          @
        STDEV IS w(r).                             @
        SPAN IS i1, i2.                            @
        OUTPUT IS LEVEL(w), PRINT(w1, w2, - - -),  @
               NOPRINT(w1, w2, - - -).             @
        HOLD RESIDUALS(v), FITTED(v), VARIANCE(v).

    Required sentence:   MODEL

    Legend:  v -- variable name;  i -- integer value;
             w -- keyword;        r -- real value
```

For more details about the different sentences and their use, you will find more information in the on-line help SCA SYSTEM SYNTAX → UNIVARIATE TIME SERIES → OESTIM

With the OFORECAST command the forecasting of time series will be carried out with adjustment in the presence of outliers. It is used like the FORECAST command. The following output was given for this paragraph

OFORECAST modelroe. NOFS are 4

```
RESIDUAL STANDARD ERROR (USES DATA UP TO THE FIRST FORECAST ORIGIN) = 5789.1


TIME     ESTIMATE    T-VALUE      TYPE
 102   18240.024       3.80        AO   ( AO: Additive Outlier, see 2.5 )
 125   16562.615       3.42        AO
 153   15209.161       2.69        AO


-----------------------------------
  4 FORECASTS, BEGINNING AT   155
-----------------------------------


TIME     FORECAST    STD. ERROR    ACTUAL IF KNOWN
 156   94161.9251    5929.6538
 157   74504.1001    6016.4412
 158   88514.5770    6101.9943
 159   99232.1413    6186.3644
```

Automatical Identification of the Model

With the command

IARIMA name of the time series

the ARIMA and SARIMA model for the series can be automatically identified and estimated. If the primary interest of the SCA session is the forecast of the time series, the model information given below, may be ignored. The IARIMA command also may include information about the seasonality of the series, if it is known. To give information about seasonality the time series should be of the type 'long series' or at least 'middle series' (see 2.2.1). Otherwise it wouldn't make sense, because the original number of measurements n is diminished to the effective number n_{eff} in dependence of the values of d, p, s, D, P (see 2.2.1). If you have data, which is monthly, you must enter 12 for the seasonality, if the season is one year.

The information about the model, which was built by the IARIMA command, is stored under the model name UTSMODEL. To get the exact maximum likelihood parameter estimates for the actual model the command

ESTIM UTSMODEL. METHOD IS EXACT

has to be entered into the system. If we enter

IARIMA ROEHM. SEASONALITY IS 12

the following information about the model will be given.

```
THE FOLLOWING ANALYSIS IS BASED ON TIME SPAN   1  THRU  155
 THE CRITICAL VALUE FOR SIGNIFICANCE TESTS OF ACF AND ESTIMATES IS 1.960

 SUMMARY FOR UNIVARIATE TIME SERIES MODEL -- UTSMODEL
 ---------------------------------------------------------------------
 VARIABLE   TYPE OF    ORIGINAL    DIFFERENCING
            VARIABLE   OR CENTERED
                                        12
 ROEHM      RANDOM     ORIGINAL    (1-B  )
 ---------------------------------------------------------------------
 PARAMETER   VARIABLE  NUM./  FACTOR  ORDER   CONS-    VALUE      STD     T
   LABEL      NAME     DENOM.                 TRAINT             ERROR   VALUE
    1                  CNST    1       0      NONE   4857.9930  264.4262  18.37
    2         ROEHM    MA      1       12     NONE    .7999      .0881    9.08
    3         ROEHM    D-AR    1       12     NONE    .3520      .1192    2.95

 TOTAL NUMBER OF OBSERVATIONS . . . .         155
 EFFECTIVE NUMBER OF OBSERVATIONS . .         131
 RESIDUAL STANDARD ERROR. . . . . . . 0.623794E+04
 --
```

Fig. 16: Results of the IARIMA Command for Time Series ROEHM

The model is of the type SARIMA(0,0,0) x (1,1,1) with s=12, P=1, D=1 and Q=1.

The seasonal MA-parameter theta has the value 0.7999 with a standard error of 0.0881 and a T-value of 9.08 (most highly significant).

The seasonal AR-parameter phi has the value 0.3520 with a standard error of 0.1192 and a T-value of 2.95 (highly significant).

The constant term has the value 4857.9930 with standard deviation 264.4262 and T = 18.37 (most highly significant).

The standard error of the residuals is $0.623794 * 10^4 = 6237.94$.

This model is of the form

$$(1-B^{12})ROEHM(t) = CNST + [(1- \Theta_1 B^{12}) / (1- \Phi_1 B^{12})]* a_t$$

To obtain the forecast for this series, the command

      FORECAST UTSMODEL. NOF number of forecasts.

has to be typed into the system. With NOF = 4 one gets the following results:

4 FORECASTS, BEGINNING AT 155

TIME	FORECAST	STD. ERROR	ACTUAL IF KNOWN
156	96665.6133	6237.8365	
157	79177.9832	6237.8365	
158	91581.6985	6237.8365	
159	102039.5355	6237.8365	

Fig. 17: Results of the command: FORECAST UTSMODEL

The commands

      - OUTLIER

      - OESTIM

      - OFORECAST

are used in the same way as before (see above *Forecasting an Estimated Model*).

The following result was given by the OUTLIER paragraph

OUTLIER utsmodel. TYPES ARE AO,IO,LS,TC

```
INITIAL RESIDUAL STANDARD ERROR =     6034.6
  TIME     ESTIMATE    T-VALUE    TYPE
   102   20185.57       3.49       IO
ADJUSTED RESIDUAL STANDARD ERROR =    5769.1
--
```

The following result was given by the OESTIM paragraph

OESTIM UTSMODEL. METHOD IS EXACT

```
THE FOLLOWING ANALYSIS IS BASED ON TIME SPAN   1   THRU   155
SUMMARY FOR UNIVARIATE TIME SERIES MODEL -- UTSMODEL
---------------------------------------------------------------------
VARIABLE    TYPE OF     ORIGINAL     DIFFERENCING
            VARIABLE    OR CENTERED
                                          12
ROEHM       RANDOM      ORIGINAL      (1-B  )
---------------------------------------------------------------------
PARAMETER   VARIABLE   NUM./  FACTOR  ORDER   CONS-    VALUE      STD      T
  LABEL      NAME      DENOM.                 TRAINT            ERROR   VALUE

   1                    CNST    1      0     NONE   4812.5895  254.8452  18.88
   2         ROEHM       MA     1      12    NONE      .8618     .0598   14.40
   3         ROEHM      D-AR    1      12    NONE      .3788     .0946    4.00


SUMMARY OF OUTLIER DETECTION AND ADJUSTMENT
------------------------------------
 TIME    ESTIMATE   T-VALUE    TYPE
------------------------------------
  102   19960.961    3.34      IO  ( IO : Innovational Outlier, see 2.5 )
------------------------------------


TOTAL NUMBER OF OBSERVATIONS. . . . . . . . . . . .       155
EFFECTIVE NUMBER OF OBSERVATIONS. . . . . . . . . . .      131
RESIDUAL STANDARD ERROR (WITHOUT OUTLIER ADJUSTMENT). .  0.621820E+04
RESIDUAL STANDARD ERROR (WITH OUTLIER ADJUSTMENT) . . .  0.596801E+04
```

The following result was given by the OFORECAST command

OFORECAST UTSMODEL.NOFS ARE 4

```
RESIDUAL STANDARD ERROR (USES DATA UP TO THE FIRST FORECAST ORIGIN)= 5968.0
```

```
TIME     ESTIMATE    T-VALUE     TYPE
 102   19960.961      3.34        IO
---------------------------------

 4 FORECASTS, BEGINNING AT  155
---------------------------------


TIME   FORECAST   STD. ERROR   ACTUAL IF KNOWN
 156   96195.2080   5968.0093
 157   79823.6294   5968.0093
 158   91516.5995   5968.0093
 159  101983.0783   5968.0093
```

During the time we worked on the present publication we found out that the IARIMA paragraph is easy to use, but will not always give the best model. It is useful to check the quality of the estimated model and if it is necessary to improve the model with the help of the TSMODEL paragraph. A way to check the quality of a model is explained in detail in chap. 5.

	IARIMA			TSMODEL		
Modeltype	SARIMA (0,0,0)x(1,1,1)			SARIMA (0,1,1)x(0,1,1)		
Parameter s	12			12		
Parameter d	0			1		
Parameter D	1			1		
	Value	Std. Err.	T-Value	Value	Std. Err.	T-Value
Parameter φ	-	-	-	-	-	-
Parameter θ	-	-	-	0.8544	0.0432	19.80
Parameter Θ	0.7999	0.0881	9.08	0.5514	0.0722	7.63
Parameter Φ	0.3520	0.1192	2.95	-	-	-
Constant	4857.9930	264.4262	18.37	-	-	-
Residual Standard Error	0.623794E+04			0.612863E+04		
Time	OForecast		Std. Err.	OForecast		Std. Err.
156	96195.2080		5968.0093	94161.9251		5929.6538
157	79823.6294		5968.0093	74504.1001		6016.4412
158	91516.5995		5968.0093	88514.5770		6101.9943
159	101983.0783		5968.0093	99232,1413		6186.3644

Table 7: Summary of the Results Obtained with the System Command Driven by Paragraphs

The improvement of the standard error of the model with the TSMODEL command is not significantly better than the model which we got from the IARIMA command. The standard deviations of the forecasts from the last three month are even smaller but also not significantly better compared with the model of the TSMODEL command. As result we can notice that both models are about of the same quality.

3.5.2   Modeling and Forecasting With the "Input Window System"

The SCA-WORKBENCH has an easy to use graphical user interface (GUI) to build time series models and make forecasts. It is organized in three primary tasks

- Automatic modeling
- User-defined modeling
- Import models from file

The tool for forecasting can be found under the menu APPS (see Fig. 7) under the option: "TS Modeling and Benchmarking". By joining this tool the following window will appear

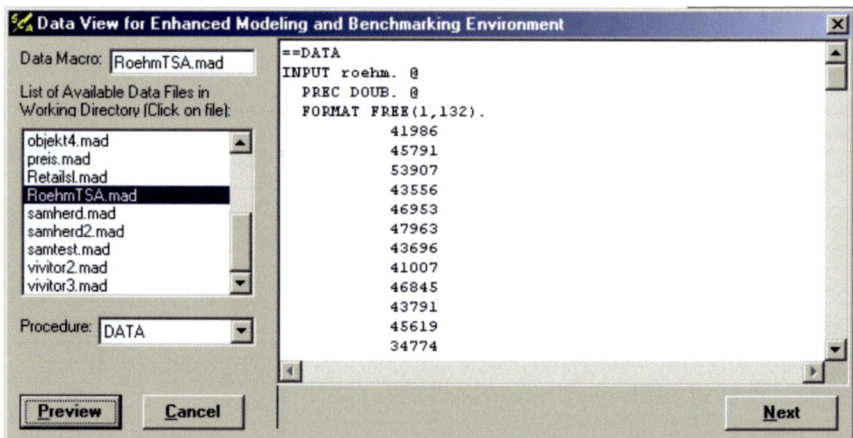

Fig. 18: Surface of Enhanced Modeling and Benchmarking

It is possible to select between all the time series, which had to be specified as SPREAD-SHEET TASKS before. How this is done is explained in 3.5.3.1.

The time series can be selected with the mouse. It will be chosen after clicking the series with the left mouse button. If wished, the series can be previewed (as it is done in the example of Fig. 18 above).

After choosing the time series, the button NEXT has to be clicked. The following window will pop up.

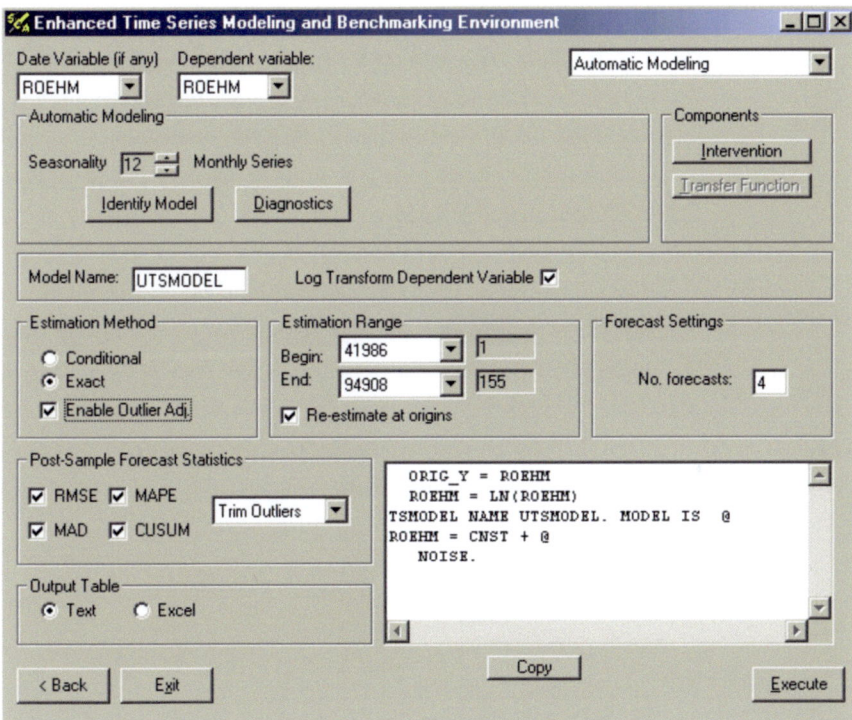

Fig. 19: The Window "Enhanced Time Series Modeling and Benchmarking"

First we shall describe the different window items.

Window Item	**Description**
Date Variable	The drop-down list may be used to specify the date variable associated with the series. If the SCA data macro contains a variable named "DATE", it is automatically assigned by SCA Workbench. Otherwise it may be selected from the drop-down list. In this case we selected ROEHM from the drop down list, but it would have been also all right, if we wouldn't have selected anything and left DATE as variable.
Dependent Variable	The drop-down list may be used to specify the time series, which has to be modeled. Independent variables may be added (i.e. transfer function components) or interventions through the *Components* frame. This variable is given by the system. So in this case the system gave ROEHM as dependent variable
Intervention	Here an intervention component can be added to the model. This item is not used in the present publication and will not be described in more detail. For more information please see [LML00] and [L00].
Transfer Function	A transfer function component might be added to the base model. This won't be discussed in the present publication. For further details see [LML00].
Model Name	The default for this field is UTSMODEL. It is possible to overwrite the model name if desired.
Log Transform	If necessary the model can be log transformed. In the case of the time series ROEHM we decided to employ the log-transformation (see 3.5.2.1), because of the results during the discussion in 3.5.1 (*Model Identification*). Note: The forecast results and confidence intervals will be transformed back into original units. For more details about log-transformation and when it is useful to employ see [No83].

Estimation Method   The method for model estimation will be selected here. If you enable outlier adjustment, the OESTIM paragraph will be used in lieu of the standard ESTIM paragraph. For more details about the settings please see [LML00].

Estimation Range   The beginning and ending span for model estimation has to be specified here. The rolling forecasting procedure will use the ending estimation span as the first forecasting origin. If the user doesn't change anything the whole series will be taken form the system. In this case we want to use the complete time series for the estimation. It might be useful to select a part of the data, if the data is not reliable in some portions.

Re-estimate at   This item should be used if the model should be re-estimated at each
Origin   forecasting origin. Otherwise, the model will be estimated once using the estimation range specified by the user. It is useful to employ this option and the system uses this tool automatically if the setting is not changed by the user.

Number of Forecasts   The user can decide how many forecasts he wants to generate. In this case we decided to have 4 forecasts for the following four months.
                       Note: The forecasts in the near future are much more reliable than forecasts which are in the late future. Therefore it makes no sense to select a high number of forecasts, for example 30.

Post-sample   Here the user can decide which performance criteria should be used in
Statistics   modeling the time series. It is advisable to employ all settings, since the result is more reliable.
                       Explanation of the items:
                       RMSE   = Root Mean Square Error
                       MAD   = Maximum Absolute Deviation
                       MAPE   = Maximum Absolute Percent Error.

In the drop down box beside the Post-sample settings the user can chose between three settings: Trim Outliers

                 Trim 5%

                 Trim none.

We used the setting "Trim Outliers" for modeling and forecasting the time series ROEHM. For more details of the outlier detection and the different settings please see [SCA01].

Output Table	Item to choose how the summary report should be produced, whether in text file format or in MS EXCEL format. MS EXCEL is required on the system to generate the reports in Excel. We decided to chose the setting "Text file" as file for the summary report.
Back	Steps back to the SCA Data Macro browser and initializes the main rolling forecasting form.
Exit	Closes the session by clicking the button
Execute	Executes the forecasting procedure and generates the forecast tables in MS EXCEL or WINDOWS-NOTEPAD.

3.5.2.1 Automatic Modeling with the "Input Window System"

The Automatic Modeling feature of the SCA-WORKBENCH is very easy to use. After the data is read into the system as explained in 3.5.1, the model has to be identified. To identify a SARIMA model, the seasonality has to be indicated. The system allows choosing between different patterns. For example, yearly seasonality with monthly data is specified as twelve, quarterly as four, and weekly seasonality with daily data as seven. If there is no seasonality specified, the Workbench will not consider a seasonal component for the model.

For the time series ROEHM we have a yearly seasonality with monthly data (see discussion in 3.5.1, *Model Identification*) and therefore we chose 12 as seasonality.

The next step in forecasting the time series will be the identification of the model. For the time series ROEHM we decided to employ the log-transformation (justification see 3.5.1, *Model Identification*). As "Estimation Method" we decided to chose EXACT and as "Estimation Range" we used the complete data from ROEHM (at all 155 points). Since it appeared useful to have 4 numbers of forecast to compare the results of the "command driven system" session with the results of the "input window session" with "automatic modeling", we set 4 as number of forecasts (see Fig. 19). Furthermore all tools for the "Post-Sample Forecast Statistics" were employed, to get a most reliable result. Since it also seemed, that there are outliers in the data of the time series ROEHM, we employed the function for *Trim Outliers*, too.

After the settings are done, the button *Identify model* has to be selected and the model equation in SCA-notation will appear in the white window at the right bottom. In the case of the above settings for the model of the time series ROEHM the following model equations appeared (see Fig. 19):

ORIG_Y = ROEHM
ROEHM = LN(ROEHM)
TSMODEL UTSMODEL. MODEL IS @
ROEHM(12) = CNST + @
(12)NOISE.

The result of identifying the model ROEHM with the tool "input window system" with "automatic modeling" is a SARIMA (p, d, q) x (P, D, Q) model of the form SARIMA (0,0,0) x (0, 1, 1).

By clicking the *Diagnostics* button, information about the residuals of the identified model will be shown in a text editor. The output includes information about the different operators of the SARIMA (p, d, q) x (P, D, Q) model and the residuals of the specified model.

```
C ***RESIDUAL DIAGNOSTICS***
   ESTIMATE UTSMODEL. HOLD RESID(RES). @
      SPAN IS 1,155.

 THE FOLLOWING ANALYSIS IS BASED ON TIME SPAN   1   THRU   155
 NONLINEAR ESTIMATION TERMINATED DUE TO:
 RELATIVE CHANGE IN (OBJECTIVE FUNCTION)**0.5 LESS THAN 0.1000D-02
 SUMMARY FOR UNIVARIATE TIME SERIES MODEL -- UTSMODEL
 -------------------------------------------------------------------

 VARIABLE    TYPE OF    ORIGINAL     DIFFERENCING
             VARIABLE   OR CENTERED
```

```
ROEHM     RANDOM     ORIGINAL     (1-B  )
---------------------------------------------------------------------

PARAMETER   VARIABLE   NUM./  FACTOR  ORDER   CONS-      VALUE     STD     T
  LABEL       NAME     DENOM.                 TRAINT             ERROR  VALUE


    1                   CNST     1      0     NONE      .0660    .0039  17.01
    2         ROEHM      MA      1     12     NONE      .5388    .0727   7.41


EFFECTIVE NUMBER OF OBSERVATIONS . .           143
R-SQUARE . . . . . . . . . . . . .           0.894
RESIDUAL STANDARD ERROR. . . . . .  0.915823E-01
--



  IDEN RES.
NAME OF THE SERIES . . . . . . . . .           RES
TIME PERIOD ANALYZED . . . . . . . . 13  TO    155
MEAN OF THE (DIFFERENCED) SERIES . . .       -0.0029
STANDARD DEVIATION OF THE SERIES . . .        0.0915
T-VALUE OF MEAN (AGAINST ZERO) . . . .       -0.3796


AUTOCORRELATIONS

  1- 12    .17  .18  .11  .16  .18  .23  .12  .20  .08 -.05  .03  .12
  ST.E.    .08  .09  .09  .09  .09  .09  .10  .10  .10  .10  .10  .10
    Q      4.2  8.8 10.6 14.6 19.5 27.7 30.0 35.9 37.0 37.4 37.5 40.0
 13- 24   -.00  .08 -.10 -.04 -.01 -.10 -.06  .03 -.05 -.22 -.03 -.22
  ST.E.    .10  .10  .10  .10  .10  .10  .11  .11  .11  .11  .11  .11
    Q     40.0 41.1 42.8 43.1 43.1 44.8 45.5 45.6 46.0 54.6 54.7 63.0


         -1.0 -0.8 -0.6 -0.4 -0.2  0.0  0.2  0.4  0.6  0.8  1.0
         +----+----+----+----+----+----+----+----+----+----+
                                    I
   1   0.17                         +    IXXXX
   2   0.18                         +    IXXXX
   3   0.11                         +    IXXX+
   4   0.16                         +    IXXXX
   5   0.18                         +    IXXX+X
   6   0.23                         +    IXXXX+X
   7   0.12                         +    IXXX +
```

```
 8    0.20                    +      IXXXXX
 9    0.08                    +      IXX   +
10   -0.05                    +      XI    +
11    0.03                    +      IX    +
12    0.12                    +      IXXX  +
13    0.00                    +      I     +
14    0.08                    +      IXX   +
15   -0.10                    +   XXXI     +
16   -0.04                    +      XI    +
17   -0.01                    +      I     +
18   -0.10                    +   XXXI     +
19   -0.06                    +      XXI   +
20    0.03                    +      IX    +
21   -0.05                    +      XI    +
22   -0.22                    X+XXXXI      +
23   -0.03                    +      XI    +
24   -0.22                    XXXXXI       +
  STOP
```

THE CURRENT SCA SESSION IS TERMINATED.

THE SIZE OF THE WORKSPACE USED IS 11004 WORDS.

Fig. 20: Results of tool DIAGNOSTIC CHECK for time series ROEHM with the "Input Window System"

After clicking the EXECUTE button (see Fig. 19), a window pops up with the information about the forecast and the model of the time series ROEHM

```
Forecast Summary Table:
Time Stamp: 16.09.2001 15:42:53
ROEHM Log-Transformed: TRUE
Estimation Method: EXACT
```

OBS/DATE	FORECAST	FCST.ERR	LCL(95%)	UCL(95%)
156.000	100227.867	9964.800	83914.523	119712.602
157.000	77854.438	7724.354	65182.645	92989.672
158.000	93700.320	9310.671	78449.414	111916.070
159.000	105980.219	10541.531	88730.602	126583.234

```
Forecasting Model:


ORIG_Y = ROEHM

ROEHM = LN(ROEHM)

TSMODEL UTSMODEL. MODEL IS @

ROEHM(12) = CNST + @

(12)NOISE.
```

Fig. 21: Results of Modeling and Forecasting the Time Series ROEHM with the "Input Window System-Automatic Modeling"

3.5.2.2   User Defined Modeling with the "Input Window System"

If the option *User defined Model* is chosen in the window of *Enhanced TS modeling and benchmarking Environment* (see Fig.18, Fig.19), the system gives you the items described in 3.5.2 and additionally 7 new items, which will be explained below.

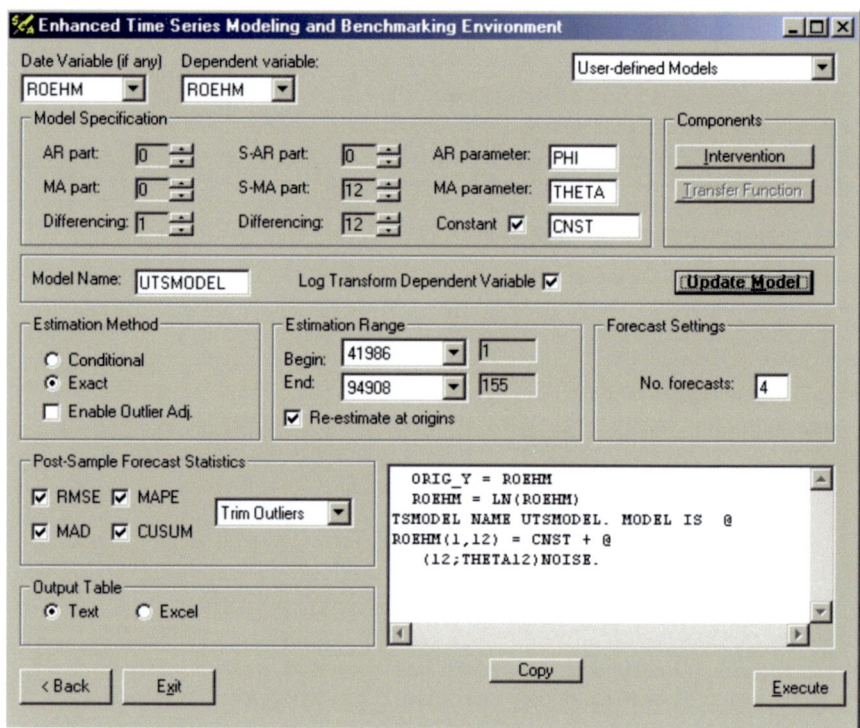

Fig. 22: Window for the "USER DEFINED" tool in the "Input window system"

The User-defined Model environment allows the user to specify time series models and have those models translated into SCA's standard syntax.

For the time series ROEHM the following settings where chosen because of the discussion in 3.5.1 (*Model Identification*).

For detailed explanations of the different parts in a SARIMA (p, d, q) x (P, D, Q) model see Appendix 1.

Menu Item	Description
AR Part	The degree of the operator polynomial for the autoregressive part of the model ROEHM is chosen as 0.
MA Part	The degree of the operator polynomial for the moving average part of the model ROEHM is also set equal to 0.
S-AR Part	The degree of the operator polynomial for the multiplicative seasonal autoregressive part of the model ROEHM is again set equal to 0.
S-MA Part	The degree of the operator polynomial for the multiplicative seasonal moving average part of the model is set equal to 12.
Differencing	The differencing orders in the model ROEHM are set equal to 1 and 12 for nonseasonal and seasonal differencing respectively according to the differencing operators (1-B) and (1-B^{12}), because of the nonstationary behaviour of the time series.
Constant	If a constant term should be included into the model it has to be mentioned here. In the case of the time series ROEHM we expect a constant term.
Update Model	After the model got changed it must be updated for SCA-Workbench. For doing so, this button has to be clicked, to construct the SCA syntax and display the information in the model text box.

The remaining options were chosen as in 3.5.2 for the automatic modeling.

The settings above are translated into the SCA language as

ORIG_Y = ROEHM

  ROEHM = LN(ROEHM)

TSMODEL NAME UTSMODEL. MODEL IS @

ROEHM(1,12) = CNST + @

  (12;THETA12)NOISE.

The following results of the model ROEHM were given by the "Input Window System" with the "user defined" option, after employing the EXECUTE button:

```
Forecast Summary Table:
Time Stamp: 16.09.2001 16:11:45
ROEHM Log-Transformed: TRUE
Estimation Method: EXACT

      OBS/DATE          FORECAST         FCST.ERR          LCL(95%)           UCL(95%)
 --------------   ---------------   ---------------   ---------------    ---------------
       156.000         86571.406         11761.625         68862.883          108833.789
       157.000         67605.438         15009.864         48912.984           93441.359
       158.000         82052.219         26070.299         55200.531          121965.617
       159.000         93342.852         40077.754         59061.234          147522.969

Forecasting Model:
  ORIG_Y = ROEHM
  ROEHM = LN(ROEHM)
TSMODEL NAME UTSMODEL. MODEL IS  @
ROEHM(1,12) = CNST + @
   (12;THETA12)NOISE.
```

Fig. 23: Results of "User Defined Modeling" in the "Input Window System"

	AUTOMATIC	**USERDEFINED**
Model type	SARIMA (0,0,0)x(0,1,1)	SARIMA (0,1,0)x(0,1,1)
Parameter s	12	12
Parameter d	0	1
Parameter D	1	1

	AUTOMATIC			USERDEFINED		
	For the transformed variable ln(ROEHM)					
	Value	Std. Err.	T-value	Value	Std. Err.	T-Value
Parameter φ	-	-	-	-	-	-
Parameter θ	-	-	-	n.o.	n.o.	n.o.
Parameter Φ	-	-	-	-	-	-
Parameter Θ	0.5388	0.0727	7.41	n.o.	n.o.	n.o.
Constant	0.0660	0.0039	17.01	-	-	-
Residual Standard Error	0.915823E-01			no output (n.o.)		
Time	For the original variable ROEHM					
	Forecast		Std. Err.	Forecast		Std. Err.
156	100227.867		9964.800	86571.406		11761.625
157	77854.438		7724.354	67605.438		15009.864
158	93700.320		9310.671	82052.219		26070.299
159	105980.219		10541.531	93342.852		40077.754

Table 8: Summary of Modeling the Time Series ROEHM with the "Input Window System"

Compared with the results of the command driven system (see 3.5.1, *Automatical Identification of the Model*), the forecast standard error is bigger, even though we employed the ln-transformation which should give a better result. Therefore it seems, that the command driven system gives better results (at least in this example), because the user has more possibilities for creating the model.

3.5.3   Modeling and Forecasting with the "System Command Driven by Macros"

In this section we will explain how a user can create macros in the SCA-System. But first we will give some information about the macro-structure.

"SCA-Workbench exploits the power of the SCA Statistical System through its use of macro procedures to automate applications" ([SCA01] page 5).

A macro procedure is a set of different SCA commands, which are created in the SCA System. These macros can call other macros and work with commands like ACF, IARIMA, OESTIM, OFORECAST, FSAVE etc.

Details of SCA macro procedures and their variable types can be found in [SCA01].

3.5.3.1   The Task Menu Item of SCA-WORKBENCH

The Task menu is a tool to create macros. The macros contain paragraphs which are used in the interactive communication with the "system command driven by paragraphs". In this section the complete Task menu is explained. The Task menu can be found in the SCA-WORK-BENCH. After selecting the option, a pop up menu appears with the following features. Those features will be explained in a summarized form. For more details see [SCA01].

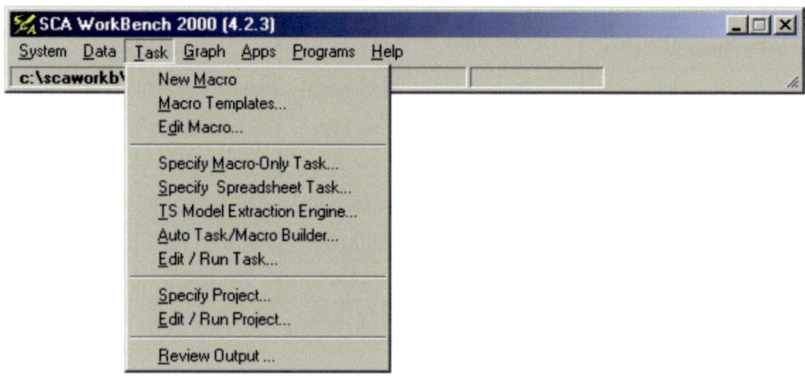

Fig. 24: The Task Menu for Creating Macros in the SCA-System

Menu	Item Description
New Macro	Opens an empty file using the default text editor, which is specified in the menu *System*, item *Environment*. For more details of these settings see [SCA01]. The user can write his new macro procedure into the editor.
Macro Templates	This item opens a generic macro-template file. The template file contains some of the basic commands of a SCA macro procedure, which may be used to modify and build macros. It is possible to add more files to this directory. We never used this feature in the present publication, but nevertheless this option may be very useful to create macros, especially if the user is only little familiar with the paragraphs in the SCA System.

Edit Macro

An existing SCA Macro or SCA Data Macro in the current working directory will be opened. We always used this option if we wanted to alter an existing data macro or add new paragraphs to an existing macro.

Specify Macro-
Only Task

This option creates a new macro-only task. This item is described in more detail below under: *The Option "Edit Macro... " of the Task Menu.* To mention is, that if a macro must be created or an existing data macro was revised, this feature has to be employed.

Specify
Spreadsheet Task

Create a new single spreadsheet task. This feature is described in detail later under: *The Option "Specify Spreadshee Task" of the Task Menu.*
If the user wants to use the automatic modeling features in the APPS menu (see Fig. 7, Fig. 24 and section 3.5.1), he also has to transform his data into a spreadsheet file.

TS Model
Extraction Engine

Tabulate estimation results from time series models that are contained within one or more SCA output files. The extraction engine supports BOX-JENKINS ARIMA, intervention, and transfer function models. Since this option was not used in the present publication it won't be described in more detail. If the reader is interested in more details he will find detailed explanations in [SCA01].

Auto Task/Macro
Builder

This item automatically generates spreadsheet tasks for a large number of data variables located in a data source. It is addressed to users who wish to apply a common analysis script for multiple variables. This feature is also suitable for batch operations and project development. Since this option was not used in the present publication it won't be described in more details. If the reader is interested in more details he will find detailed explanations in [LML00].

Edit/Run Task	Edits or executes an existing SCA Macro-only Task or Spreadsheet Task. This option is also included in the option *Specify Macro Only Task* (see above).

Specify Project   Creates a new project that consists of any number of tasks, which will run in a sequential manner. It allows to group several tasks together. Once the tasks are grouped into a project, SCA-WORKBENCH uses a *.PRJ extension for project file names. This option was not used in the present publication. If the reader is interested in details, he will find more explanations in [SCA01].

Edit/Run Project   This item edits or executes an existing project. Clicking this feature shows all project files that are located in the working directory. The same run options are available as in the options *Specify Project* and *Specify Spreadsheet Task* (see above), which is explained in the following section. The option *Edit/Run Project* was not used in the present publication. If the reader is interested in details, he will find more explanations in [SCA01].

Review Output   This feature opens an SCA output file. The user can select the output file from an existing macro which ran already by employing the option *RUN* in the feature *Specify Macro-Only Task* (see above).

The most important options in the TASK menu, which will be explained below, are
- Edit macro
- Specify Macro-only Task
- Specify Spreadsheet Task
- Edit/Run Task
- Review Output.

3.5.3.2   Important Options of the Task Menu

The first step in creating a macro procedure is to write a complete new procedure in the editor
which opens after employing the option *New Macro*, or first to create a *SPREADSHEET
TASK* from existing data. The existing data may be available as an EXCEL sheet or an ASCII
file. Since it was easier to create a new macro, when the data input already has been accomplished, we always first employed the *Spreadsheet Task* option, before writing the commands
into the macro procedure.

<u>The Option *Specify Spreadsheet Task* of the Task Menu</u>

To create a new single spreadsheet task the item *Specify Spreadsheet Task* has to be clicked in
the TASK menu. This will activate a new dialogue box and list all available spreadsheet files
located in the working directory. If the spreadsheet is located in another directory, it has to be
selected by navigating through the system. But the files should be located in the working directory to avoid any problems with the portability of the new spreadsheet.

Fig. 25: The Window of the Option *Specify Spreadsheet* Task of SCA-Workbench

After choosing the file ROEHMTSA.XLS and clicking the OPEN button the task specification dialogue box will appear.

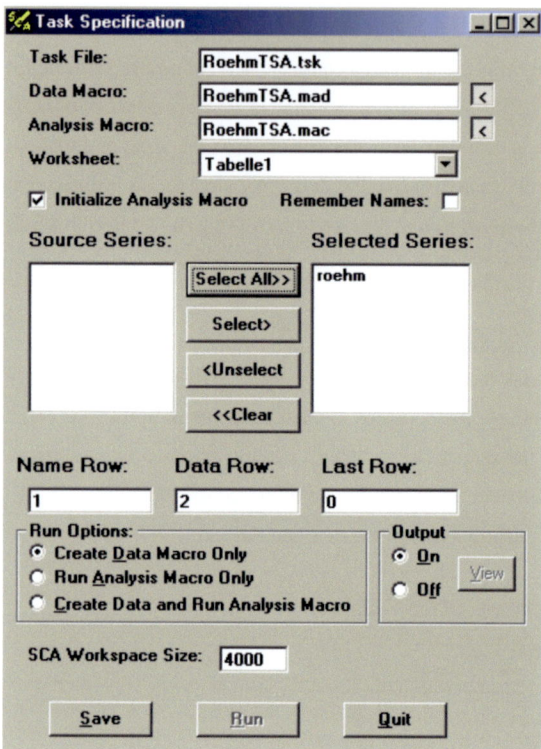

Fig. 26: Specification of the Spreadsheet Task for the Time Series ROEHM

SCA WorkBench requires that the worksheet is in columnar format and that the series names are located in the first row. This format is typically referred to as a standard database format. If there is a worksheet that is not in standard database format, the system can still interactively build an SCA data macro by using the mouse to highlight the cells in the worksheet and by using the "Save Data As" option in the Data menu of SCA Workbench. This is explained in detail in [SCA01].

We will give a short explanation of the different settings. For more details see [SCA01].

Item	Item Description
Task File	The task file inherits the name of the worksheet but substitutes its extension by *.TSK. It may be overwritten if desired.

Data Macro   The data macro file inherits the name of the worksheet but substitutes its extension by *.MAD. It also can be overwritten if desired. The actual data macro file is not generated until the task is saved and run. The data macro for time series ROEHM is of the following form

     = DATA
     INPUT roehm. @
      PREC DOUB. @
      FORMAT FREE(1,132).
         41986
         45791
         53907
    ...
         98639
         94908
     END OF DATA
     -- This data macro was created on: 11.06.2001 16:23:23
     -- Number of data rows written: 155
     RETURN

Analysis Macro   The first level analysis macro file inherits the name of the worksheet but substitutes its extension by *.MAC. If needed the name has to be overwritten.

Worksheet   Displays the name of the worksheet. The default selection is the first worksheet in the spreadsheet file. If the worksheet is multiple, an arrow will appear in the rightmost part of the box to specify a different worksheet. If a new worksheet is selected, the variable information in the list boxes will be updated automatically.

Initialize Macro   Initializes an analysis macro file that links a specified data macro to the analysis macro in the first level macro.

Remember Names	If a number of related tasks is created that will have similar names, SCA Workbench may be wished to use the last task file, data macro, and analysis macro file names entered instead of the default names that are tied to the worksheet name. By clicking on the *Remember Names* check box this option will be activated for the remainder of the session.
Source Series	This list box shows all series located in the worksheet presented in alphabetical order. The worksheet must be organized in columnar format where the first row contains the series names and the corresponding data are located in the same column below the series name.
Select All	By clicking on the *Select All* command button all source series will be moved to the selected series list box. As a rule, not more than 10 should be selected. The SCA System has a limit of 132 characters per line. If 132 characters are exceeded in the generated data macro, the data may not be read into the SCA System correctly.
Select	By highlighting a source series and clicking on the *Select* command button the highlighted series will be moved to the *Selected Series* list box.
Unselect	Works similar to the *Select* command button but in the opposite direction. In other words, it moves series from the *Selected Series* list box to the *Source Series* list box.
Clear	Moves all series in the *Selected Series* list box back to the *Source series* list box.
Name Row	Specifies the row in the worksheet that contains the series names.

This setting is currently fixed to "1" and cannot be changed. The series names must be located in the first row of the worksheet for this capability to function correctly.

Data Row   Specifies the beginning row for the data. This feature allows to skip rows at the top of the spreadsheet column. The default is row 2.

Last Row   Specifies the last row for the data. If the default "0" is used as the last row, it will include all available data located between the beginning row specified in Data Row and the bottom of the worksheet. The end of data marker is generated in the worksheet when an entire data row is blank.

Run Options   After saving a task, one may execute the task from this dialog box. There are given three choices (1) Create Data Macro Only, (2) Run Analysis Macro Only, or (3) Create Data and Run Analysis Macro. If attempting to run an analysis task in which the analysis macro or the data macro is missing; an error message is printed in the *Track Window* and the task is aborted.

Output   If this item is selected to save the output for an analysis, the default is on. The output is saved to a *.OUT file.

SCA Workspace   Specifies the workspace size for the PC SCA System.

<u>Structure of the Spreadsheet Task for the Time Series ROEHM:</u>

```
== ALLMACRO
 ASSIGN FILE 12. EXTERNAL 'RoehmTSA.mad'.
 CALL DATA. FILE 12.
C Time series variableROEHM may be referenced by positional variable: &V_1.
 CALL ANALYSIS
 STOP
```

= ANALYSIS

C

C Insert your commands here

C

 RETURN

The Option "Edit Macro…" of the Task Menu

With the option EDIT MACRO, an existing macro procedure can be opened. The user just has to click on "Edit Macro…". A window will appear with the following form.

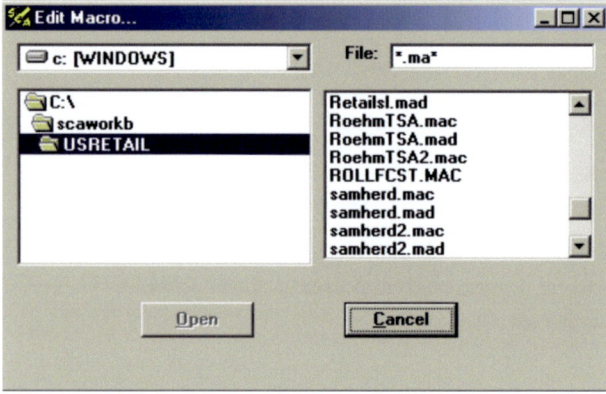

Fig. 27: Edit Macro Window of the Task Menu

Now the user can choose which macro shall be edited.

Note: Every procedure appears two times, one time with the extension .MAD and the second time with the extension .MAC. In the . "mad-file" the user will find the data of the time series, for which he had created a spreadsheet task before. The "mac-file" contains the paragraphs and instructions of the macro-procedure.

Example for time series ROEHM:

The "mad-file" RoehmTSA.mad is of the following form:

== DATA

INPUT roehm. @

```
PREC DOUB. @
FORMAT FREE(1,132).
      41986
      45791
      53907
      ...
      99357
      98639
      94908
END OF DATA
-- This data macro was created on: 11.06.2001 16:23:23
-- Number of data rows written: 155
RETURN
```

Regarding the "mac-file" RoehmTSA.mac the following structure will be given.

```
==ALLMACRO
  ASSIGN FILE 12. EXTERNAL 'RoehmTSA.mad'.
  CALL DATA. FILE 12.
C Time series variable ROEHM may be referenced by the positional variable: &V_1.
  CALL ANALYSIS
  STOP
==ANALYSIS
C
C  Insert your commands here
C
  RETURN
```

Note: The CALL DATA paragraph is the link to the data-file "RoehmTSA.mad", which reads the data for the time series into the actual SCA-session.

The user will insert his specific instructions into the "mac-file". Those instructions are used in the same way as in the "system command driven by paragraphs". The difference is, that the user can list all commands ("paragraphs") and the complete results of the different steps of the analysis will be given in the output-file (see Appendix 2).

For the time series ROEHM we created the following macro procedure for analyzing the data. The explanations to each command are written in comment lines beginning with the letter "C". The reasons, why this structure was created can be found in chap. 4 (*Working With Time Series in SCAGRAF*) and chap. 5.

```
==ALLMACRO
   ASSIGN FILE 12. EXTERNAL 'ROEHMTSA.MAD'.
C ROEHMTSA.MAD IS THE FILE WHICH CONTAINS THE DATA OF THE TIME
C SERIES

CALL DATA. FILE 12.
C READING THE DATA OF  TIME SERIES ROEHM INTO THE SCA-SYSTEM
C THE VARIABLE ROEHM MAY BE REFERENCED BY POSITIONAL VARIABLE: &V_1.

CALL ANALYSIS
   STOP
==ANALYSIS

C AUTOMATIC MODELING
IARIMA &V_1.SEASONALITY IS 12
C AUTOMATIC MODELING OF THE DATA

ACF &V_1. MAXLAG IS 36
C DEPICTING OF THE ACF FOR THE TIME SERIES ROEHM

OUTLIER UTSMODEL. TYPES ARE AO,IO,LS,TC
C OUTLIER DETECTION OF THE MODEL CREATED BY THE IARIMA COMMAND

OESTIM UTSMODEL. METHOD IS EXACT
C ESTIMATION OF THE MODELPARAMETERS WITH OUTLIER ADJUSTMENT

NEW SERIES IN V1, V2, V3.
C SAVES THE INFORMATION DURING THE OUTLIER DETECTION PROCESS
C V1: NAME USED TO STORE RESIDUALS AFTER OUTLIER ADUSTMENT
```

```
C V2: NAME FOR THE ADJUSTED SERIES
C V3: NAME FOR THE INDICATOR VARIABLE WITH THE TYPE OF THE
C OUTLIER.
C V1 AND V2 WILL BE NEEDED FOR THE SCAGRAF SESSION TO PRINT
C THE OUTLIER PLOT

FORECAST UTSMODEL. NOFS ARE 4
C FORECAST FOR TIME SERIES MODEL FOUND BY THE IARIMA COMMAND
OFORECAST UTSMODEL.NOFS ARE 4
C FORECAST INCLUDING THE OUTLIER INFORMATION

C USER DEFINED MODELING
TSMODEL NAME IS MODELROE. MODEL IS &V_1(1,12)=(1-THETA1*B)@
(1-THETA12*B**12)NOISE
C EXPECTED MODEL TYPE BASED ON ANALYSES OF GRAPHIC DEPICTING
C AND FORMER SCA SESSIONS

ESTIM MODELROE. METHOD IS EXACT. HOLD RESIDUALS(RESR)
C MODEL ESTIMATING
ACF RESR. MAXLAG IS 36
C DEPICTING OF THE AUTOCORELATIONS OF THE RESIDUALS OF TIME SERIES
C ROEHM WITH THE EXPECTED MODEL

OUTLIER MODELROE. TYPES ARE AO,IO,LS,TC
C OUTLIER DETECTION OF THE USER DEFINED MODEL

OESTIM MODELROE. METHOD IS EXACT
C PARAMETER ESTIMATION AND ADJUSTING THE USER DEFINED MODEL TO
C OUTLIERS

NEW SERIES IN X1, X2, X3.
C SAVES THE INFORMATION DURING THE OUTLIER DETECTION PROCESS
C X1: NAME USED TO STORE RESIDUALS AFTER OUTLIER ADUSTMENT
C X2: NAME FOR THE ADJUSTED SERIES
C X3: NAME FOR THE INDICATOR VARIABLE WITH THE TYPE OF THE
C OUTLIER.
C X1 AND X2 WILL BE NEEDED FOR THE SCAGRAF SESSION TO PRINT
C THE OUTLIER PLOT
```

```
FORECAST MODELROE.NOFS ARE 4
C FORECAST FOR THE USER DEFINED MODEL


OFORECAST MODELROE. NOFS ARE 4
C FORECAST FOR OUTLIER ADJUSTED USER DEFINED MODEL


RETURN
```

After the existing macro-file is revised, it must be saved. For saving the menu FILE in the WINDOWS Editor NOTEPAD must be used. Under this menu the user will find the option SAVE. The next step is to run the revised macro. This can be done in the option *Specify Macro-Only Task* (see below) or with the option *Edit/Run Task...* (see below).

The results generated by the macro for the time series ROEHM are listed in Appendix 2.

The Option "Specify Macro-only Task" of the Task Menu

"A macro-only task places a wrapper around a self-contained SCA macro procedure so a task name can be associated with it. This allows SCA Workbench to process self-contained macros in the standard mode of a task and group these tasks into projects" [SCA01]. To specify a macro-only task, the user has to click on an individual SCA macro procedure which is displayed in the file list box. The task file that is associated with the self-contained macro will have the same name as the macro file selected, but its file extension will be *.TSK. By clicking the *Save* button the task file will be created in the current working directory. It is advised that the default name of the macro-only task is not modified, because of associating the SCA macro with the task file it belongs to.

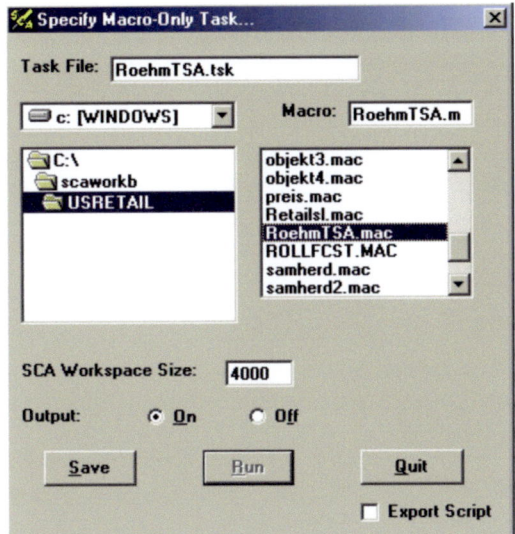

Fig. 28: Window for the Option "Macro-Only Task" of the Task Menu

After the new or revised macro procedure is saved, it is recommended to run the macro by clicking the *Run*-button. This is useful because then the new output file for the specific macro is saved as output-file and can be opened with the *Review output* (see below) option of the TASK menu.

The Option "Edit/Run" of the Task Menu

After a macro-only task or a spreadsheet task is specified, the individual task will be edited or run by choosing the *Edit/Run* option. The *Edit/Run Task* menu item shows all task files that are located in the working directory (see Fig. 29). For editing a task located outside of the current working directory, it is recommended that the working directory be changed by using the option *System Profile* of the System menu. Otherwise, the changes will be saved in the working directory and it may cause path conflict errors. By clicking on the *Edit* command button an existing task file will be edited. A dialogue box will be displayed that is very similar to the *Specify Spreadsheet Task* dialogue box.

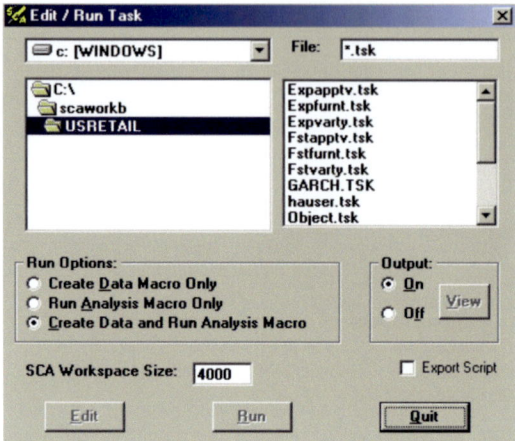

Fig. 29: The Window of the Option "Edit/Run" of the Task Menu.

The Option "Review Output" of the Task Menu

This option depicts the result of the SCA session with the specific macro. After choosing this option a window will appear, where the user can choose among the different output files. For the example of the time series ROEHM you will find the output file in the Appendix 2.

4   Graphical Depicting of Time Series with the SCAGRAF Tool

To depict data in different forms is an essential tool in the analysis of a data set. Especially for time series analyses, it is the first step in forecasting the series. The time plot will give sensible information about transformation, trends, seasonality and possible outliers.

The SCA System has a number of commands for displaying time plots, which will be discussed in this chapter.

To start the SCAGRAF System, it has to be selected in the WORKBENCH at the menu *GRAPH* with the option: *Run SCAGRAF*.

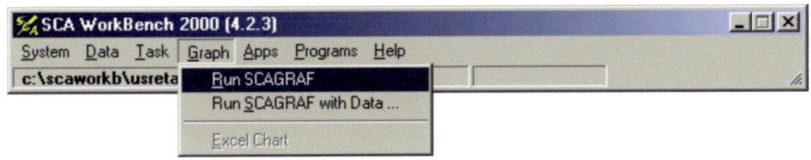

Fig. 30: Menu Item for SCAGRAF

After opening SCAGRAF the menu bar will appear on the screen. A short description of the different menu options and their functions is listed below.

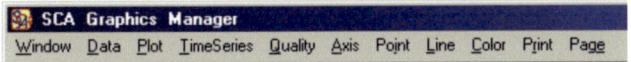

Fig. 31: SCAGRAF Menu

Menu Item	Description
Window	For opening and rearranging the layout of graphics windows.
Data	Working with data in the workspace, like reading, writing, editing, creating, renaming and deleting.
Plot	Creates plots for time series, scatter and contour plots and for changing the size of windows and graphs.

Time Series   Tool for the visual analysis of time series with instruments for aggrega-
                   tion, transformation, differencing and depicting the ACF. Additionally
                   it provides forecast plots, outlier plots and simple regression capabili-
                   ties.

Quality   Used for generating plots in quality control.

Axis   For manipulating shape, text, labels and graph legends of the axis

Point   For changing shape and size of the points in the graph

Line   For changing thickness and shape of lines in the plot

Color   For changing the color of points, lines, axes, text and background of the
                   graph

Print   For printing the graph

Page   For displaying the graphic windows by every screen.

Working with files in SCAGRAF

Before the SCAGRAF plotting capability can be used, the specified data has to be stored into
the workspace of the system. The data may be in an ASCII file or a SCA formatted file, which
was created during a SCA session.

To store the data into the workspace, the menu DATA has to be clicked and then the specific
format of the data has to be selected, for example ASCII- formatted (see Fig. 32). A dialogue
box will appear, with the available data in this format. Double clicking may choose this data.
If an ASCII file has no variable name in the first row, a window will pop up and ask for a
variable, which may be entered and then confirmed by clicking *ok*. Another possibility is, to
read a SCA formatted file into the workspace of SCAGRAF. This file must have been created
and saved during a SCA session. One example, when this is useful, is when the user wants to
print the outliers of a time series. The user has to save the specific data of the OESTIM para-

graph as explained in section 3.5.1. Those files, which are created and saved during an SCA session have the ending FMT. To enter other formatted data, the option is very similar to the ASCII option.

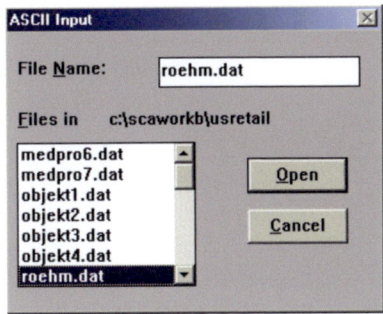

Fig 32: Input Window for ASCII Data in SCAGRAF

Plot of a Time Series

For a simple plot of the time series, the user has to choose the item PLOT. This item will open a submenu, were the user can find the option "SINGLE SERIES PLOT..." (option number 8). After choosing this option a window will pop up, where the time series must be selected which should be printed. By confirming the settings the plot of the time series will appear on the screen.

Working With Time Series in SCAGRAF

There are different capabilities for time series analysis and forecasting. The available tools in the menu TIME SERIES are

- Aggregation
- Transformation
- Differencing data
- Computing the ACF
- Forecast plot
- Outlier plot

Aggregation of Time Series in SCAGRAF

Aggregation means to smooth the time series. It is easier to recognize trends, if the time series is aggregated instead of in its original form. For aggregating time series, the *Aggregation* item has to be selected under the menu *Time Series*. A dialogue box of the following form will appear.

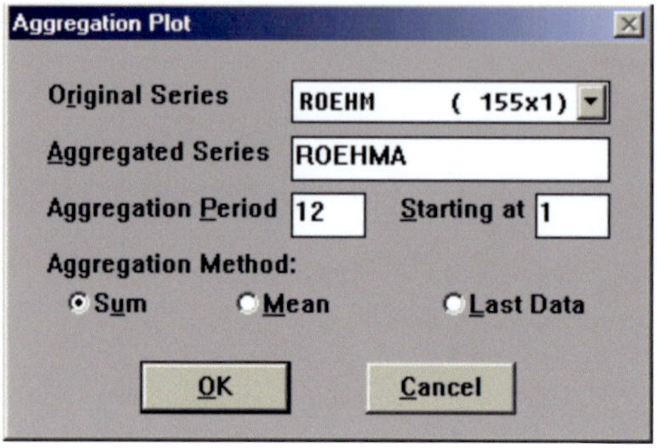

Fig. 33: Aggregation Window

The variable name of the original time series has to be selected in the first text box. For the aggregated series a name has to be chosen. The next box will set the aggregation period. For example: For generating yearly data out of monthly data the number 12 has to be entered. The starting point for the aggregation has to be selected in the next box and at last the method for the aggregation has to be chosen.

Sum   adds together all observations of an aggregation period

Mean   computes the average of the observations

Last Data   uses the last observation of an aggregation period

Confirm your settings with the *ok* Button. An example is given for the time series ROEHM. The original series is displayed before the aggregated series.

You will get the graphs of the series and the aggregated series by choosing

WORKBENCH→ *GRAPH*→ RUN SCAGRAF →PLOT →SINGLE SERIES PLOT
WORKBENCH→ *GRAPH*→ RUN SCAGRAF →TIME SERIES → AGGREGATION.
The results for the time series ROEHM are shown in Fig. 34 and Fig. 35.

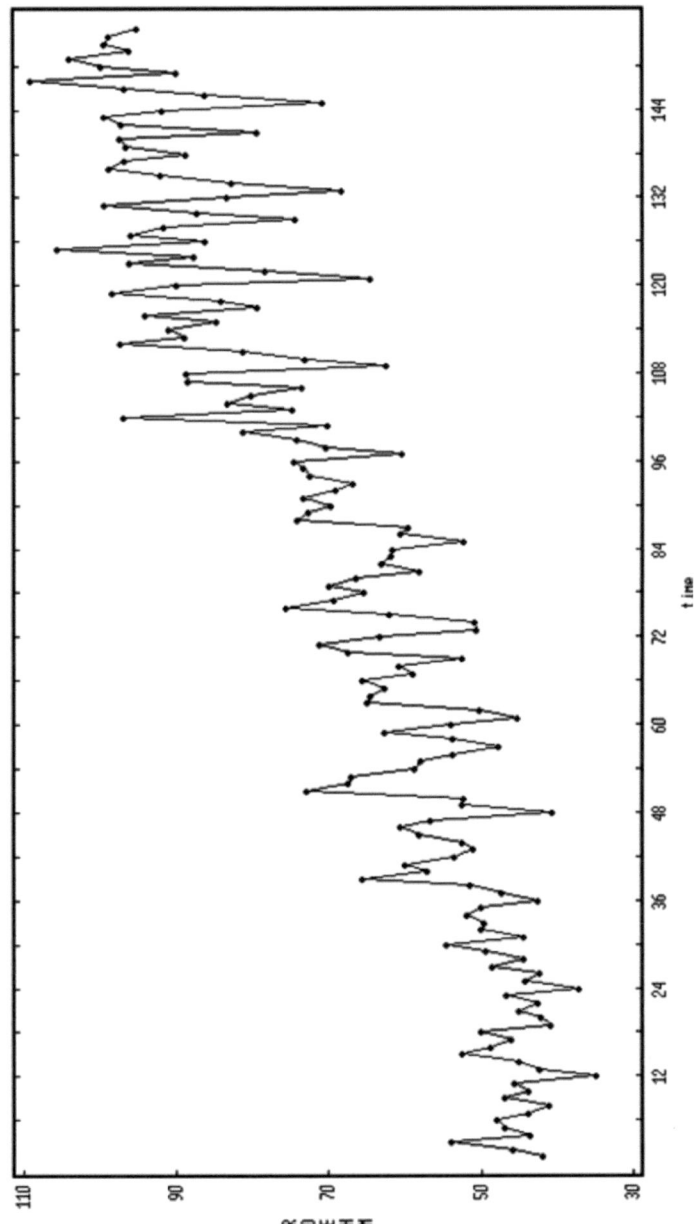

Fig. 34: Plot of Time Series ROEHM

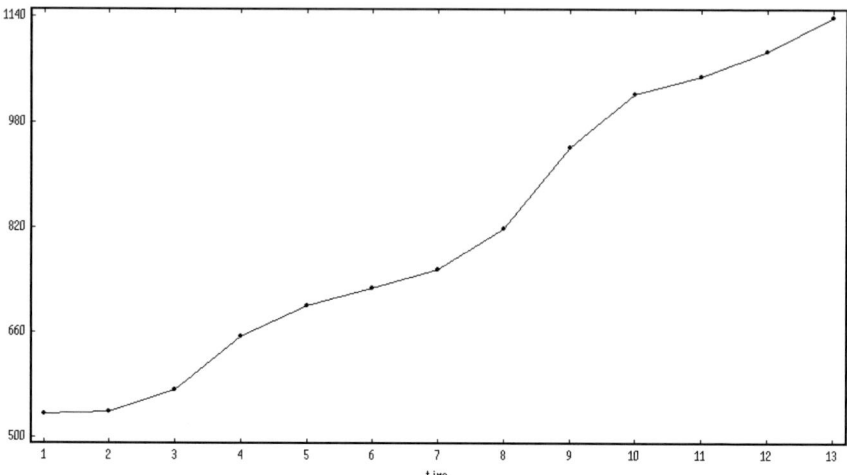

Fig. 35: Aggregated Plot of Time Series ROEHM

The settings which where chosen in the AGGREGATION window: monthly data into yearly data with the aggregation method *sum.*

Transformation of Time Series with SCAGRAF
Time series can be analysed by using an ARIMA/SARIMA model. A data transformation which is useful in many cases is the BOX-COX transformation. Details about this kind of transformation will be found in [G00]. More details about the option TRANSFORMATION of the SCAGRAF menu *Time Series* will be found in [SCA95]. If the graph shows changes like increasing spreading, then this is a hint for nonstationary variances. In this case a **ln**-transformation is useful.

Differencing Time Series in SCAGRAF
The differencing operation is another form of transforming data. It works with subtraction of the value of a prior observation from the current observation. This item is located in the menu TIME SERIES below the transforming item. Under DIFFERENCING ORDERS the user may enter how often differencing should be carried out. If differencing order = 2, then the differences are differenced again. The settings from the system are:
the first order difference (1),

a seasonal difference (12),

and the first difference of the seasonal difference (1,12).

(Note: This can be changed by the user).

Those orders are useful in monthly data. The orders have to be separated by a semicolon. As an example the differenced time series ROEHM with d = 1 is depicted below.

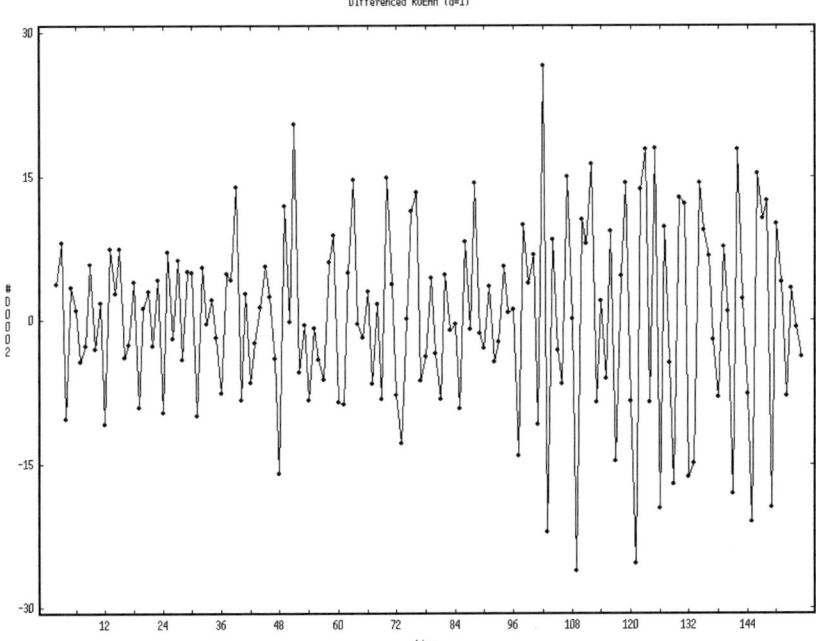

Fig. 36: Differenced Time Series ROEHM with d = 1

Computing the ACF in SCAGRAF

The meaning of the ACF was already explained in section 2.4.1. The tool for computing the ACF is located in the menu TIME SERIES (see Fig. 31). The maximum lag has to be set by the user, the default maximum lag is 36. If the ACF should be computed for various differencing orders of the original series, this may be entered into the control label *Differencing Order(s)*. The orders are separated by comma.

SCA- convention	Meaning
1	represents (1-B)
2	represents (1-B)^2
12	represents (1-B^{12})
1,12	represents (1-B) (1-B^{12}).

If the 95% confidence interval for the ACF should be displayed, the YES- Button has to be selected. For choosing just a part of the series for the correlation a *Data Span* has to be selected in the specified box. An example of a computed ACF is shown below.

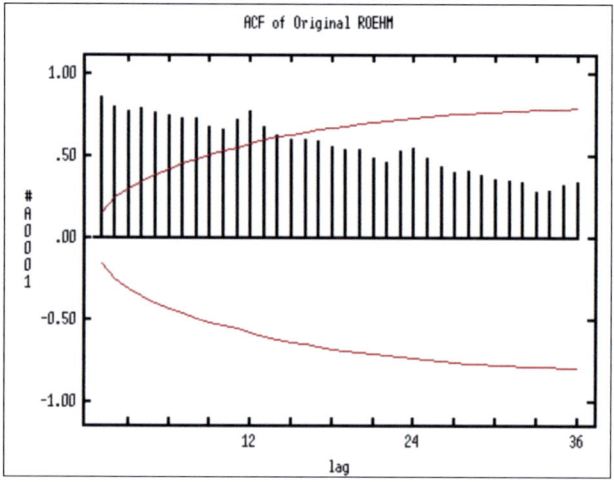

Fig. 37: Graph of the ACF of the Time Series ROEHM in SCAGRAF

The ACF of the time series ROEHM has large values and decays very slowly. This behavior is typical of a non-stationary series and indicates that we should difference the series. Obviously there is also a seasonality included. This is recognizable through the "waves" of the ACF. Every twelfth line in the ACF is a local peak. The first twelve lines are above the significance level. Differencing of the order one with d = 1 has to be done to get the TS stationary. Because of the seasonality it may be advisable to have also a differencing of the type (1-B ^{12}).

For detailed information how to use the ACF in modeling time series see [No83], [G00] and [LML05].

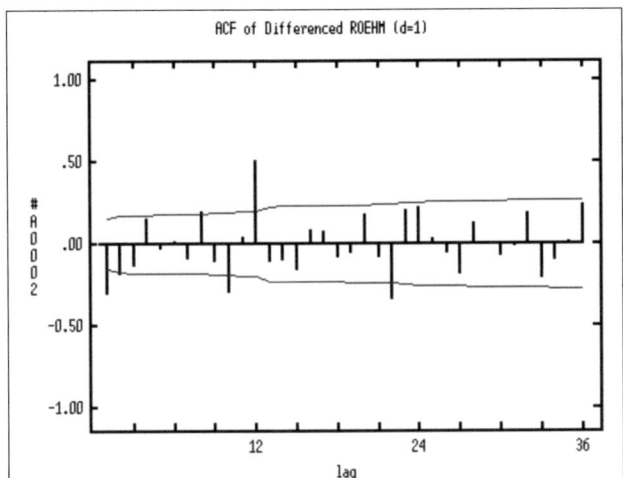

Fig. 38: ACF of the Differenced Time Series ROEHM with Nonseasonal Differencing (d = 1)

Clearly the use of (1-B) alone does not remove the effects of non-stationarity from the data since the ACF at lags 10, 12, 22 is great. Seasonal differencing is warranted. It seems that a non-seasonal and a seasonal differencing operator of the form $(1-B)*(1-B^{12})$ must be employed. The corresponding ACF is depicted below.

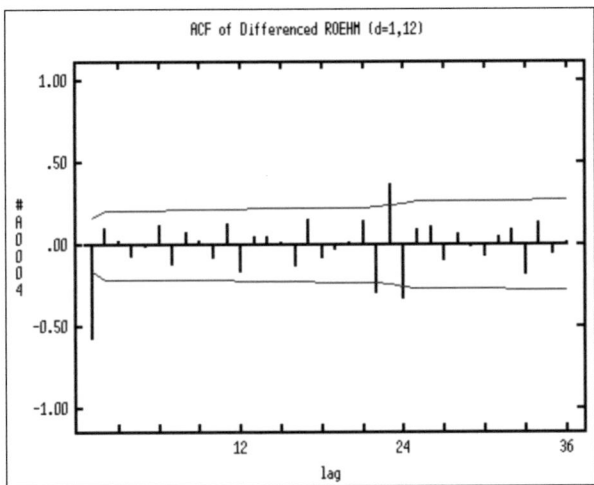

Fig. 39: ACF of the Differenced Time Series ROEHM with Nonseasonal and Seasonal Differencing (d = 1, s = 12, D = 1)

We may assume that the model of the Time Series ROEHM will be of the type SARIMA (0,1,0) x (1,1,1) with the seasonality s = 12. This information will be needed in section 3.5.1 for the specification of the model and in section 3.5.3.1 for the macro procedure.

Forecast Plot of Time Series in SCAGRAF

SCAGRAF provides graphical capabilities for plotting forecasts with their confidence intervals, but it does not compute forecasts. To create a forecast plot the data for the forecasts and the data for the confidence intervals have to be read into the workspace of SCAGRAF. Needed are:

-    the original time series
-    the forecast
-    the standard errors of the forecast.

Forecasting the time series with SCA EXPERT or the WORKBENCH will deliver this information. Once the data is stored in the workspace the forecast plot can be created. The option for plotting the time series with the forecasts and their confidence intervals is called FORECAST PLOT and contained in the menu TIME SERIES (see Fig. 31).

The following dialogue box will appear:

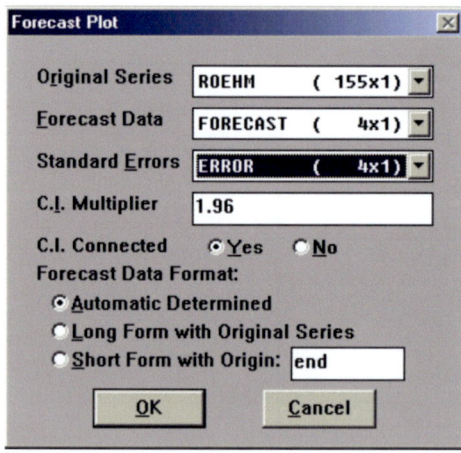

Fig. 40: Forecast Plot Window

A plot of the time series ROEHM with the forecasts is given in Fig. 41 on the following page.

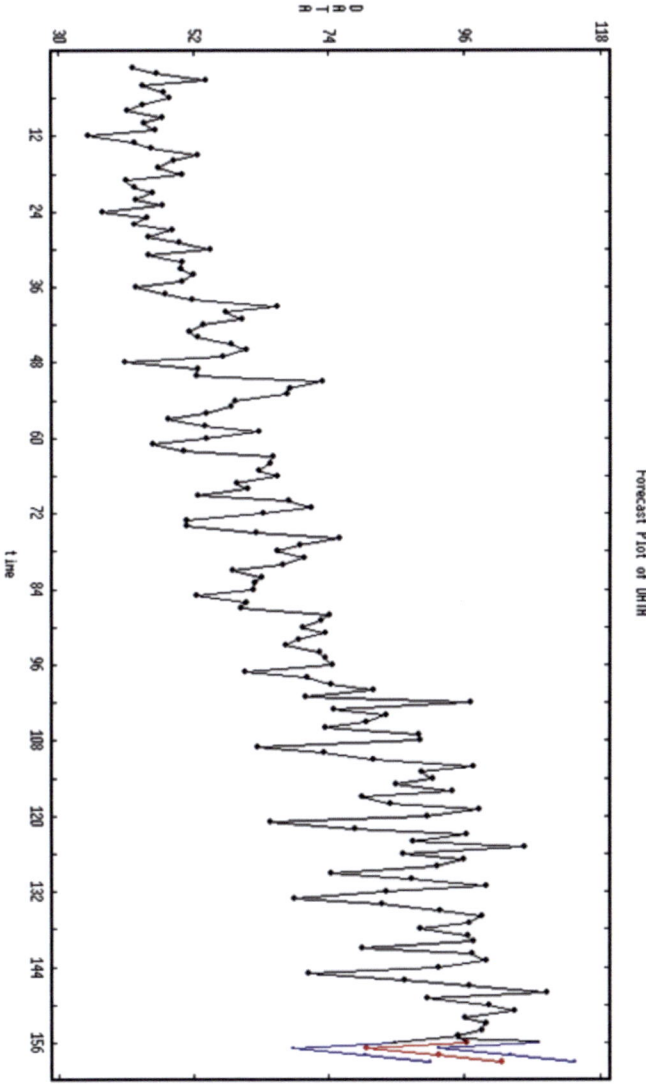

Fig. 41: Forecast Plot of Time Series ROEHM

Annotation: <u>Calculation of the Confidence Interval (CI)</u>

0. Confidence Intervals (CI) With Non-transformed Measurements:

f_t: forecast of the measurements x_t

e_t: forecast error for f_t

   $CI = f_t \pm c(S) * e_t$

1. Confidence Intervals (CI)for Ln- transformed Measurements :

g_t: forecast of transformed measurements $y_t : = \ln (x_t)$

h_t: forecast error for g_t

CI for transformed measurements y_t:

   $CI = g_t \pm c(S) * h_t$

CI for original measurements $x_t = \exp(y_t)$:

   $CI = \exp(g_t \pm c(S) * h_t) = \exp(g_t)* \exp(\pm c(S) * h_t)$

with the upper sign (+) for the upper confidence limit and the lower sign (-) for the lower confidence limit.

After choosing the specific data the C.I. Multiplier has to be set. For a confidence interval of safety probability $S = 95\%$ the value is 1.96.

Safety probability S [%]	C.I. Multiplier c(S)
98	2.326
95	1.960
90	1.645
80	1.282
70	1.036

If the points, which comprise the confidence interval, should be connected, the button YES in the Forecast Plot Window of Fig.40 has to be selected. There are also three different options for the format of the graph for the forecast data.

Long form: the forecasts are appended to the original series and displayed as one series

Short form: only the forecasts are displayed as a separate series

Automatic determined: SCAGRAF determines which format is used

For illustrating a forecast plot with the time series ROEHM, the option "Long form" was taken.

The data for the forecast plot were taken from the SCA session before (see section 3.5.1: *Forecasting an Estimated Model*).

Outlier Plot of Time Series in SCAGRAF

To properly model a time series, outliers have to be adjusted. To create an outlier plot in SCAGRAF the adjusted time series has to be read into the workspace of the system. The adjusted series is a series with the effect of the outliers removed. The indicator series is a series that is as big as the adjusted series and holds the information which type of outlier appears at the specific points of the time series. Each element of the indicator series specifies whether an observation has been classified as an outlier and its type. The codes for this indicator series are

0   No outlier
2   Innovational Outlier IO
3   Additive Outlier AO
4   Temporary Change TC
5   Level Shift LS

The explanation for the different types can be found in section 2.5.

To get the adjusted series and the indicator series, we used the OESTIM paragraph (command) in the SCA session before with the specific sentences (parameters) as it was explained in section 3.5.1: *Forecasting an Estimated Model*.

To illustrate the outlier plot, we used the time series ROEHM again. The original series is connected, while the adjusted series is plotted as points. At the 102. and the 125. observation appear outliers of the type AO in the original series. As data for the outlier plot the results of the session with the commands TSMODEL and OESTIM in the Appendix 2 were used.

Note: The disadvantage of SCAGRAF (in the version we used) is, that the quality of the plots is not good enough to read the exact values out of the plots. It would be easier, if lines would be included. After the plot is printed on paper it is even harder to recognize the values of the specific observations.

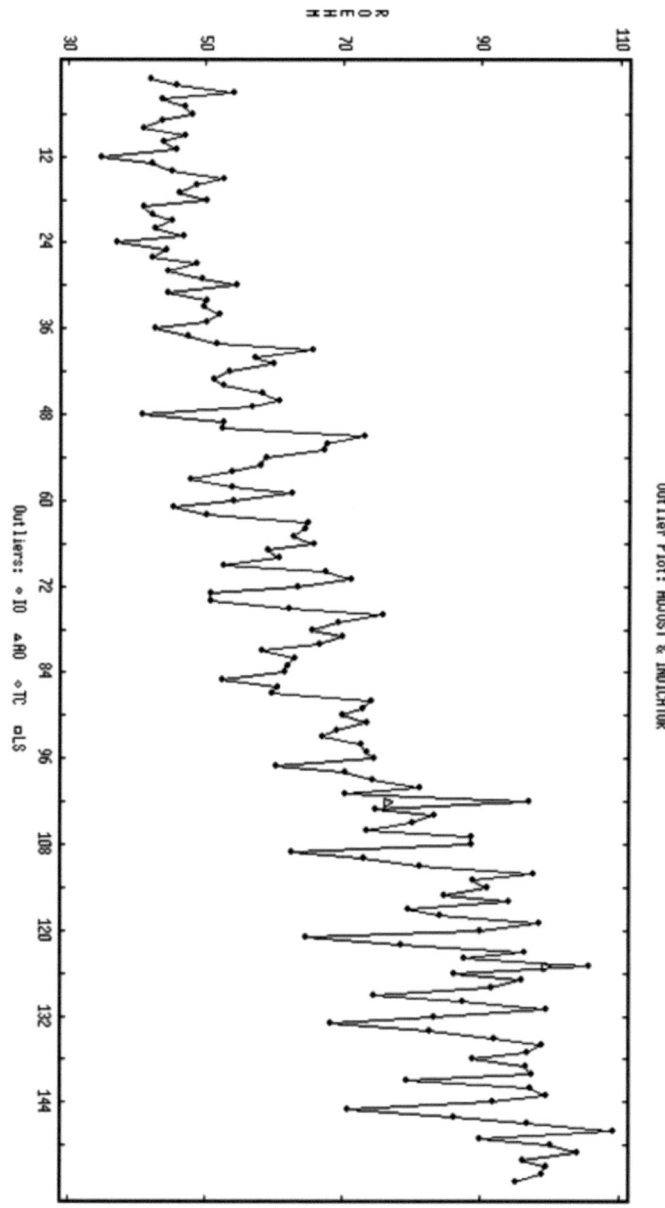

Fig. 42: Outlier Plot of Time Series ROEHM

5   Recommendations for Working with the SCA-System

This chapter will give some general recommendations for the usage of the SCA-System in the order of the Input- Process-Output (IPO) principle.

First we will give some general information of the working manner of the SCA-System. Both versions, the SCA 32-bit version (system command driven by paragraphs) and the SCA WORKBENCH (input window system) are just surfaces. In the back it is just one system. The WORKBENCH is a new surface, which makes it easy to use the SCA 32-bit system. The WORKBENCH activates the SCA 32-bit version with the help of macros. The user doesn't need to plug in the paragraphs (i. e. commands) in an interactive manner. The third possibility is a mixture of both. The user works with the help of windows to create macros (system command driven by macros).

If the user decides to model the specific time series in an automatic way, he may join the *APPS* menu (see Fig. 7) and click on *TS-modeling and Benchmarking* and then join the feature *automatic modeling*. In the back of the system the IARIMA paragraph gets called by a macro procedure. The macro procedure includes all the settings which where chosen in the window of the *automatic modeling* feature. For example the OUTLIER paragraph will be employed if *outlier trim* got activated. The process is executed in the SCA 32-bit system which sends the results back to the WORKBENCH, where the user can view the results with the "Execute" button. There are just two different ways to model and forecast time series: Automatic - with the IARIMA command and user defined - with the TSMODEL command. Below you will find a table which summarizes the different options.

	Automatic	**User defined**
32-bit version – command	IARIMA	TSMODEL
WORKBENCH/ APPS menu option	AUTOMATIC- MODELING	USER DEFINED- MODELING

Table 9: Overview over the Different Options for Times Series Modeling with the SCA-System

Note: In the USER DEFINED-MODELING only degrees of the AR- and MA-operator polynomials can be used, that means p, q for the non-seasonal and P, Q for the seasonal time series model. A polynomial with gaps cannot be employed.

In the WORKBENCH the user also has the chance to create macros in an easy way of its own. With this option it is possible to include *Automatic modeling* and *User defined modeling* in one macro process. The user just hast to employ the *TASK* menu (see Fig. 7). He may decide which paragraphs he wants to have employed. He can model the time series automatically and user defined and then compare which model is of better quality. This is the recommended way of forecasting time series and will be explained in detail in the following sections.

5.1   Recommendation for Data Input

The recommendation for the data input is to create a *Spreadsheet task* with the specific data. This is useful because the WORKBENCH creates a macro procedure, which inputs the data. The workflow for creating a *Spreadsheet task* was explained with the example of the time series ROEHM from section 3.5.3.1.

5.2   Recommendation for Data Processing

For processing the data we recommend to create a macro procedure which models and forecasts the time series in two different ways:

- an automatic manner
- an user defined manner

In the end, both parts will be compared and the solution with the better, which is the smaller, AIC value will be taken. The AKAIKE Information Criterion (AIC) is defined as

$$AIC := \ln s^2 + k* \ln (N)/N = 2 \ln s + k* \ln (N)/N$$

s = residual standard error of the model

k = number of parameters

N = number of measurements.

Note: The values for s and k will be created during the run of the macro procedure. Their values can be found in the output file

The results for the 4 different AIC for the time series ROEHM (see Appendix 2) are:

AIC1 = 2 ln 6218.20 + 3(ln(155)/155) = 17,57

for the automatic modeling without outlier adjustment

AIC2 = 2 ln 5968.01 + 3(ln(155)/155) = 17,49

for the automatic modeling with outlier adjustment

AIC3 = 2 ln 6450.85 + 2(ln(155)/155) = 17,61

for the user defined modeling without outlier adjustment

AIC4 = 2 ln 5929.96 + 2(ln(155)/155) = 17,44

for the user defined modeling with outlier adjustment.

The macro procedure will be created in the task menu, after a spreadsheet task of the specific data was made. You will open the spreadsheet task with the option EDIT MACRO in the TASK menu. After selecting the specific spreadsheet task a window will open with a macro procedure for the data input. It will be of the following form:

```
ASSIGN FILE 12. EXTERNAL 'ROEHM.mad'.
CALL DATA. FILE 12.
CALL ANALYSIS
STOP
==ANALYSIS
C
C Insert your commands here
C
RETURN
```

The details, how the file ROEHM.mad looks, can be found in section 3.5.3.1 : *The Option Specify Spreadsheet Task of the Task Menu.*

For the automatic part of modeling you will employ the paragraphs (commands)

- IARIMA
- OUTLIER
- OESTIM
- OFORECAST.

For the user defined part you will employ the following paragraphs (commands):

- TSMODEL
- OUTLIER
- OESTIM
- OFORECAST.

The information about the expected model for the TSMODEL paragraph will be found with employing the tool SCAGRAF option *Time Series*. With the help of the ACF you will get information about possible model types. The usage of those tools is explained in section 2.4. The recommended macro procedure for the time series ROEHM is depicted in section 3.5.3.1: *Structure of the Spreadsheet Task for the Time Series ROEHM* and *The Option "Edit Macro" of the Task Menu*.

5.3   Recommendation for the Data Output

If the user worked with a macro procedure, the best way of the data output is to employ the option REVIEW OUTPUT in the TASK menu. The commands and results of the whole macro procedure will be saved in a NOTEPAD file. This feature is comparable to the REVIEW command. The output file for the time series ROEHM is listed in the Appendix 2.

5.4   Recommendation for the Standard Report

The standard report should include the model with the best (i. e. smallest) AIC value. In the case of the time series ROEHM we will include the model we received by user defined modeling with outlier adjustment, because the AIC of this model has the smallest value (see section 5.2).

Since it depends on who the reader of the standard report is, we suggest to distinguish between three types of standard reports.

We decided to distinguish between the standard report for

- Experts
- Management
- Top-Management.

The content of the standard report for experts should be organized in three main parts

- Basic information about the time series
- Model and forecasts of the time series
- Plots of the forecasted time series.

The content of the standard report for the management should be organized in two main parts

- Model and forecasts of the time series
- Plots of the forecasted time series.

For the top-management just one part is interesting and this is the

- Forecasts of the time series.

The information about the model and forecasts for the standard report are taken form the data output file which is shown by employing the TASK menu option "REVIEW OUTPUT". The different plots are created with SCAGRAF as explained in chap. 4.

Basic Information About the Time Series

The basic information of the time series should include the data itself and a plot of the time series. The data should be copied from the file which contains the values of the time series into a MS-WORD-document with *copy* and *paste*.

The time series plot will be taken from the SCAGRAF-session. To copy those plots into the MS-WORD document it is recommended to import the plot into a graphic program. We used the PAINTBRUSH software, which is included in MS-WINDOWS. An easy way to transfer the plot into PAINTBRUSH is, to employ the "Print"-Key and the "Insert" option after opening PAINTBRUSH. The actual screen is automatically transferred into the actual document of PAINTBRUSH. Now the user can select the part he wants to copy into the MS-WORD document and transfer it with *copy* and *paste*.

Model and Forecasts of the Time Series

The information of the model and the forecasts of the time series will be found in the output file of the specific macro procedure. For the time series ROEHM this information can be found in the Appendix 2. You also will find the forecast figures of the SCA-session in the Appendix 2. The information will be again transferred with *copy* and *paste* into the MS-WORD document.

The transformation of the model from SCA-notation into the polynomial form is explained in the Appendix 1.

A more elegant way is to employ the SCA applet, with which the user can transfer the data directly into the specific document. More details about this method can be found in the online help of the SCA-system.

Plots of the Forecasted Time Series

The forecast plot and the outlier plot are created in SCAGRAF (see chap. 4: *Forecast Plot of Time Series in SCAGRAF* and *Outlier Plot of Time Series in SCAGRAF*) and transferred into the MS-WORD document as explained above for the time series plot.

The standard report of time series ROEHM for an expert will be found below and on the following pages. In section 6.1 you will find an example of a standard report for the management with using the time series SAM1. The last example is the time series SAM2 and its standard report is created for the top management. You will find it in section 6.2.

Since the quality of the outlier plot is not good, we suggest to mark the outliers with a circle, after it was transferred to a graphic program (see above under *Basic Information About the Time Series*) and give a short description, what kind of outlier appears at this point.

THE STANDARD REPORT FOR EXPERTS FOR THE TIME SERIES ROEHM
Basic Information of the Time Series ROEHM
The data of the series ROEHM

41986 45791 53907 43556 46953 47963 43696 41007 46848 43791 45619 34774 42258 45003 52523 48653 46137 50070
40930 42157 45118 42477 46713 37055 44186 42259 48506 44404 49475 54469 44505 50010 49674 51855 50061 42521
47337 51527 65477 57144 59917 53437 1097 52473 58098 60569 56631 40642 52584 52381 72881 67493 67044 58751
57951 53819 47682 53722 62563 53950 45169 50142 64842 64415 62627 65591 59014 60736 52459 67380 71128 63357
50581 50733 62097 75513 69218 65436 69893 66436 58176 62963 61890 61543 52331 60538 5961373991 72627 69765
73305 69039 66818 72441 73264 74370 60276 70232 74129 81001 70159 96764 74636 83122 80009 73393 88407 88584
62441 72983 81014 97336 88744 90728 84642 93958 79324 83911 98252 89841 64508 78214 96092 87518 105516 85963
95749 91359 74336 87154 99349 83082 68199 82508 91902 98661 96598 88627 96332 97254 79244 97116 99335 91661
70662 86012 96677 109153 89718 99822 103851 95995 99357 98639 94908

Number of measurements n= 155

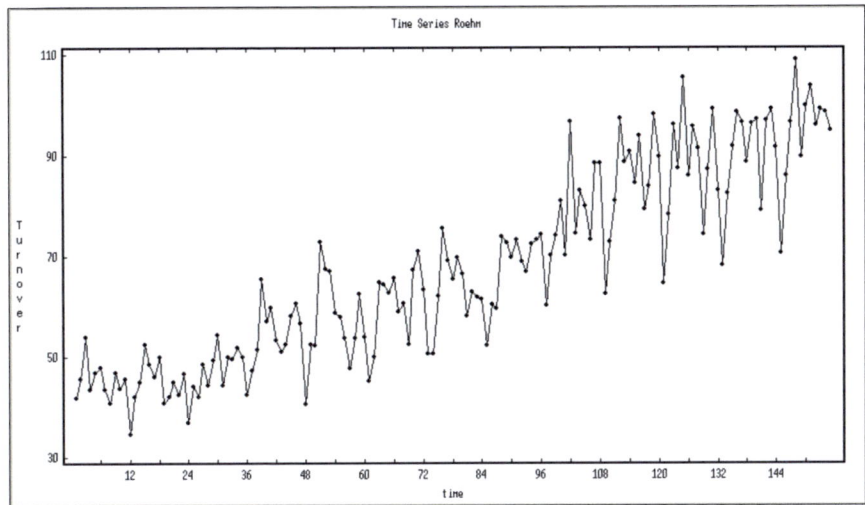

The Plot of the Time Series ROEHM

Note: The values for the turnover must be multiplied with 1000

The ACF of the Time Series ROEHM

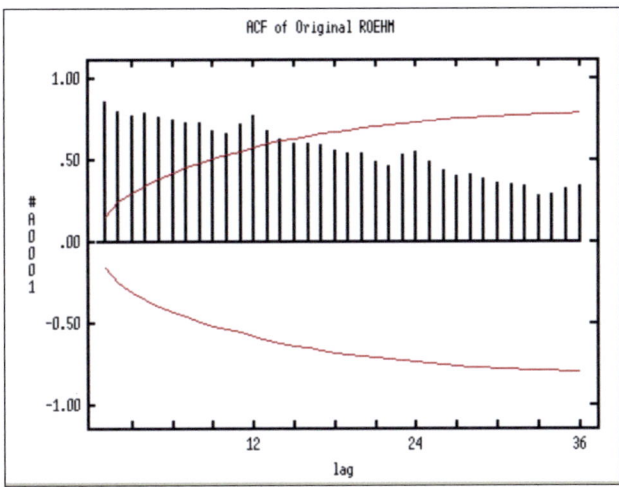

The ACF for the Original Series ROEHM

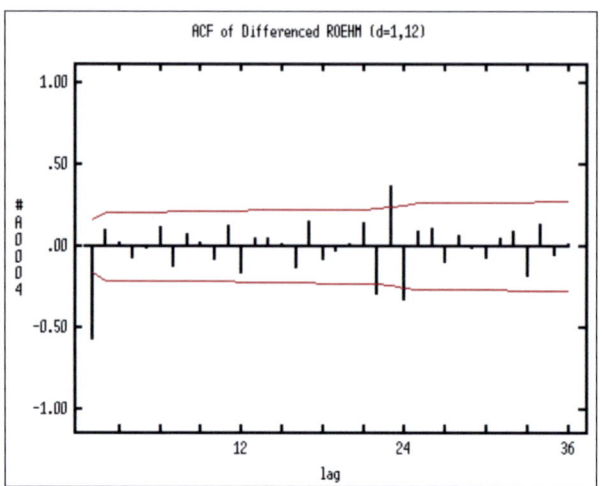

The ACF of the Differenced Series ROEHM

The difference orders (d = 1,12) mean nonseasonal differencing of order 1 and seasonal differencing with seasonality s = 12.

Model Information of the Time Series ROEHM

The time series ROEHM is of the form SARIMA (0,1,1) x (0,1,1)

In operator form this will be

$(1\text{-}B)\,(1\text{-}B^{12})\,y_t = (1\text{-}(0.8283\,;0.0488)^*B) * (1\text{-}(0.5125\,;0.0773)^* B^{12}) * a_t$

The numbers in brackets separated by colon (;) are the parameters and their standard errors.

```
                                             1         12
    ROEHM       RANDOM      ORIGINAL      (1-B  )  (1-B  )
    -------------------------------------------------------------------

    PARAMETER   VARIABLE NUM./ FACTOR ORDER CONS-    VALUE   STD     T      SIGNIFI-
    LABEL       NAME     DENOM.                TRAINT         ERROR   VALUE  CANCE

    1 THETA1    ROEHM    MA      1       1     NONE   .8283   .0488   16.97  ****
    2 THETA12   ROEHM    MA      2      12     NONE   .5125   .0773   6.63   ****
```

Outlier Figures of the Model ROEHM

```
TIME    ESTIMATE   T-VALUE    TYPE     SIGNIFICANCE

102   18.313927     3.72       AO      ****  most highly significant
125   16.790133     3.39       AO      ****  most highly significant
```

Forecast Figures of the Model ROEHM

```
TIME    FORECAST   STD. ERROR         LOWER LIMIT       UPPER LIMIT

156   94161.9251   5929.6538 (6.29%)  88232.2713       100091.5789
157   74504.1001   6016.4412 (8.07%)  68487.6589        80520.5413
158   88514.5770   6101.9943 (6.89%)  82412.5827        94616.5713
159   99232.1413   6186.3644 (6.23%)  93045.7769       105418.5057
```

Outlier Plot of the Time Series ROEHM

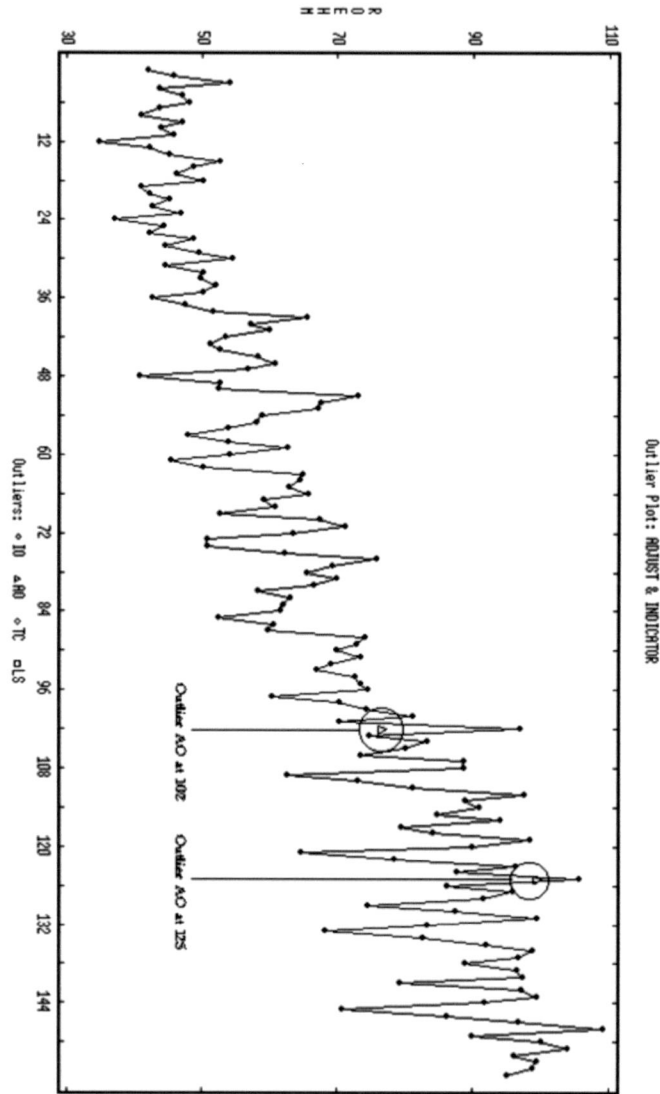

Forecast Plot of the Time Series ROEHM

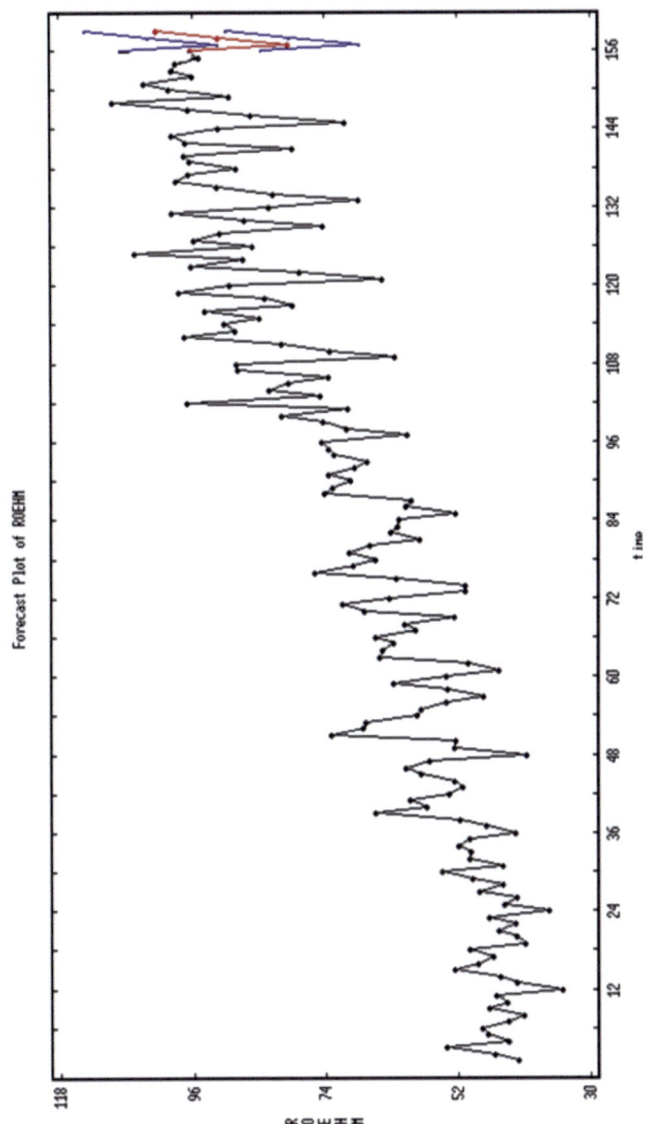

The SCA-System includes an additional tool for the output of the results into an MS-EXCEL-sheet. This feature is called EXCEL-APPLET. In this feature the user may choose which information he wants to transfer to the MS-EXCEL sheet. More information about this additional tool can be found in the online help of SCA in *Applet program help files*.

6 Case Studies with the SCA-System for Regular Time Series in the CHNL

This chapter will focus on the analyses and forecasts of two examples of regular time series. The data of the examples are taken from the waste section. The results of forecasting these time series with the BOX-JENKINS method were used to benefit the decision for future actions in this issue.

6.1 Forecasting Time Series SAM1

The data are given by the monthly measurements of waste in Rheinland- Pfalz, which needs to be watched because of its danger. The data consists of the different time series SAM1 and SAM2. The values of the time series SAM1 are listed below. SAM1 will be analyzed and forecasted using the recommended macro for the time series ROEHM in section 3.5.3.1 (see *Structure of the Spreadsheet Task for the Time Series ROEHM*).

SAM1

24293 20137 24527 23883 26053 24422 23717 17756 21063 24020 20868 19616 19869 19227 20179
22728 17252 14800 16108 19236 16673 19812 19095 22189 12506 16136 13876 15571 16970 19673
15775 12918 18313 16810 16689 16322 20221 20379 24244 14883 42820 26542 23101 23120 31006
32779 33618 28822 18672 19177 32345 25367 30566 35542 45467 39213 38986 46243 48578 34923
40678 41027 52723 53860 43044 48852 47252 47247 41747 35492 28353 46582 20467 33218 39375
45857 34169 44386 51620 44168 44523 46992 54935 53734 51140 33607 75473 47366 46623 47390
44261 54624 49274 66861 48833 49886 35300 41712 51252 44115 43592 45510 51442 44571 67351
70141 69070 61841 56749 82140 75391 72115 65619

Number of measurements n = 113

Fig. 43: Time Series SAM1

The data is saved as ASCII file in the working directory as SAMHERD.dat. It must be read from left to right, line by line. You will find the data macro in Appendix 3.

Depicting the Time Series SAM1 with SCAGRAF

(You will find detailed information for the depicting process in chap.4)

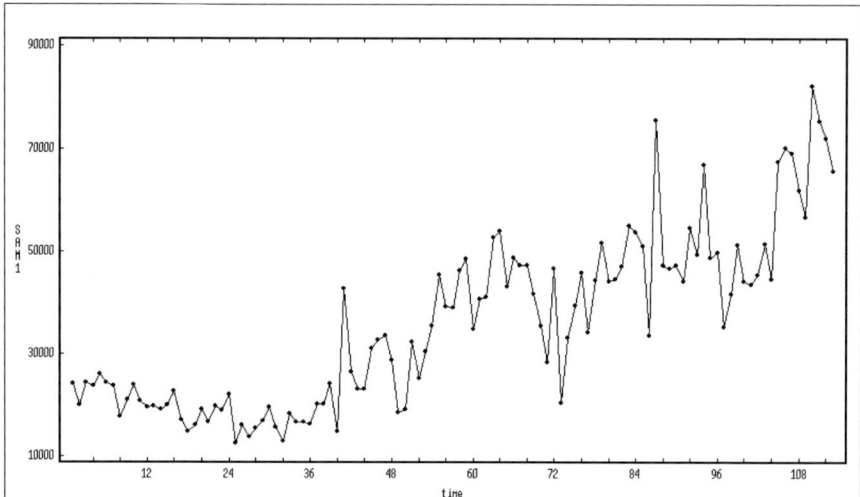

Fig. 44: Plot of Time Series SAM1

In this it is hard to say if the time series SAM1 follows a linear trend. The time series is not stationary. We assume that the time series follows a non linear trend. Since the spreading increases with time a log-transformation might be useful but is not used in the present analysis, because the result was also very good without the ln-transformation. It is obvious that there are several outliers in the data.

It may be useful to employ the aggregation option from SCAGRAF (see chap.4: *Aggregation of Time Series in SCAGRAF*). Because the time series SAM1 is monthly data it seems to be useful to set as period for aggregation the value twelve. Again as method SUM was chosen.

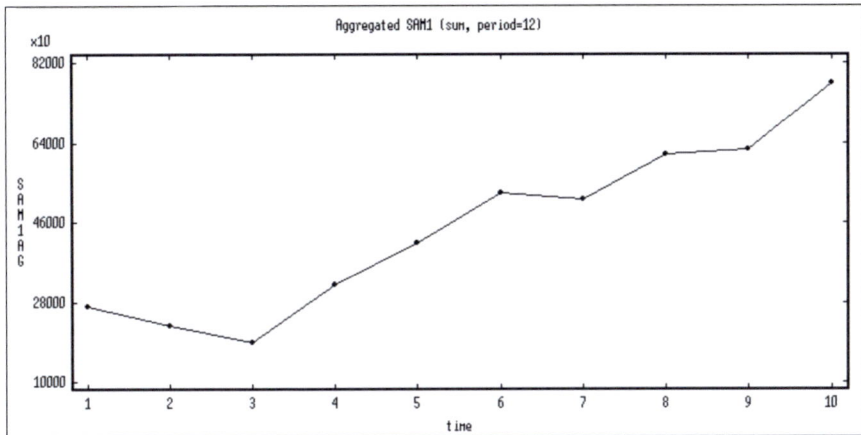

Fig. 45: Plot of Aggregated Time Series SAM1

It seems, that the time series first follows a negative linear trend and then it changes into a positive linear trend. The reason for this reaction may be the appearance of an outlier. The outlier seems to be of the type Level Shift LS. Because of the trends differencing will be warranted. As next step the ACF of the time series SAM1 must be plotted (for details how this works see chap.4: *Working with Time Series in SCAGRAF*).

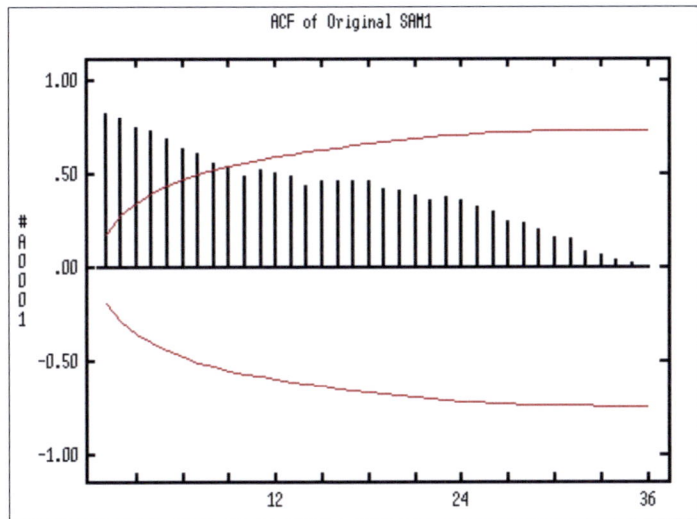

Fig. 46: ACF of the Time Series SAM1

The ACF of the time series SAM1 has large values and decays slowly. This is again a secure sign for a non-stationary series and the hint for differencing the data. Obviously there is no seasonality included because the local peaks follow no regular pattern. The first 8 values are above the significance limit. More information in the ACF about the model type will be found after the series is differenced with the operator (1-B).

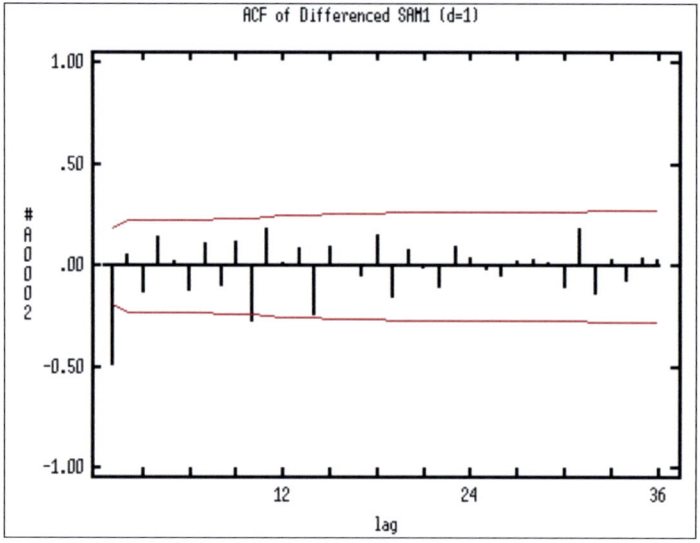

Fig. 47: ACF of the Differenced Time Series SAM1 with d = 1

It seems, that the use of nonseasonal differencing of order d = 1 does remove the effects of the trend form the data. The ACF of the differenced series dies out, this allows us to assume that there is a MA-part in the model. We tentatively use an ARIMA (0,1,1)-model for the time series SAM1.

We now will forecast the time series SAM1 with the help of a macro procedure. This procedure is of the following form:

Note: You will find the macro DATA and the file SAMHERD.MAD in the Appendix 3.

```
==ALLMACRO
  ASSIGN FILE 12. EXTERNAL 'SAMHERD.MAD'.
```

```
   CALL DATA. FILE 12.
C TIME SERIES VARIABLE SERIE1 MAY BE REFERENCED BY POSITIONAL
C VARIABLE: &V_1.
   CALL ANALYSIS
   STOP
==ANALYSIS
C AUTOMATIC MODELING
IARIMA &V_1.SEASONALITY IS 12
C AUTOMATIC MODELING OF THE DATA


ACF &V_1. MAXLAG IS 36
C DEPICTING OF THE ACF FOR THE TIME SERIES SAM1


OUTLIER UTSMODEL. TYPES ARE AO,IO,LS,TC
C OUTLIER DETECTION OF THE MODEL CREATED BY THE IARIMA COMMAND


OESTIM UTSMODEL. METHOD IS EXACT
C ESTIMATION  OF THE MODEL PARAMETERS INCLUDING THE
C INFORMATION OF THE OUTLIERS


     NEW SERIES IN V1, V2, V3.
C SAVES THE INFORMATION DURING THE OUTLIER DETECTION PROCESS
C V1: NAME USED TO STORE RESIDUALS AFTER OUTLIER ADJUSTMENT
C V2: NAME FOR THE ADJUSTED SERIES
C V3: NAME FOR THE INDICATOR VARIABLE WITH THE TYPE OF THE
C OUTLIERS.
C V2 AND V3 WILL BE NEEDED FOR THE SCAGRAF SESSION TO PRINT
C THE OUTLIER PLOT


FORECAST UTSMODEL. NOFS ARE 4
C FORECAST FOR TIME SERIES FOUND BY IARIMA


OFORECAST UTSMODEL. NOFS ARE 4
C FORECAST FIGURES INCLUDING THE OUTLIER INFORMATION
```

```
C USER DEFINED MODELING


TSMODEL NAME IS MODELSAM. MODEL IS &V_1(1)=(1-THETA1*B)NOISE
C EXPECTED MODEL TYPE BASED ON ANALYSIS OF GRAPHIC DEPICTING


ESTIM MODELSAM. METHOD IS EXACT. HOLD RESIDUALS(RESR)
C MODEL CREATING


ACF RESR. MAXLAG IS 36
C DEPICTING OF RESIDUALS OF TIME SERIES SAM1 WITH THE
C EXPECTED MODEL


OUTLIER MODELSAM. TYPES ARE AO,IO,LS,TC
C OUTLIER DETECTION OF THE USER DEFINED MODEL


OESTIM MODELSAM. METHOD IS EXACT
C ADJUSTING THE USER DEFINED MODEL TO OUTLIERS


NEW SERIES IN X1, X2, X3.
C SAVES THE INFORMATION DURING THE OUTLIER DETECTION PROCESS
C X1: NAME USED TO STORE RESIDUALS AFTER OUTLIER ADJUSTMENT
C X2: NAME FOR THE ADJUSTED SERIES
C X3: NAME FOR THE INDICATOR VARIABLES WITH THE TYPE OF THE
C OUTLIER.
C X2 AND X3 WILL BE NEEDED FOR THE SCAGRAF SESSION TO PRINT
C THE OUTLIER PLOT


FORECAST MODELSAM. NOFS ARE 4
C FORECAST OF THE USER DEFINED MODEL


OFORECAST MODELSAM. NOFS ARE 4
C FORECAST FOR OUTLIER ADJUSTED USER DEFINED MODEL
RETURN
```

The values for the 4 different AICs (see Appendix 3: *Output for Time Seies SAM1* (Residual Standard Error)) are:

AIC1 = 2 ln7872.54 + 1* ln(113)/113 = 18.01: Automatic modeling without outlier adjustment

AIC2 = 2 ln4314.69 + 1* ln(113)/113 = 16.80: Automatic modeling with outlier adjustment

AIC3 = 2 ln7871.68 + 1*ln(113)/113 = 18.01: User defined modeling without outlier adjustment

AIC4 = 2 ln4314.69 + 1* ln(113)/113 = 16.80: User defined modeling with outlier adjustment

AIC2 = AIC4 is the smallest value. Therefore we will take the results with outlier adjustment for the standard report. The standard report of the time series SAM1 in the form of the report for managers is given below and on the following pages. You will find the output of the macro procedure attached in the Appendix 3.

STANDARD REPORT FOR MANAGERS FOR THE TIME SERIES SAM1

The following "waste-figures" are expected for the next four month. This forecast is based on the analysis of the time series SAM1 with the BOX-JENKINS method.

Month 01	65791.2587	(± 6.55%)
Month 02	64374.9937	(± 7.13%)
Month 03	63383.6082	(± 7.66%)
Month 04	62689.6384	(± 8.13%).

The values in brackets are the standard deviations of the forecasts.

The model for the times series is of the form ARIMA(0,1,1) with nonseasonal differencing of order d = 1 and moving average parameter Theta = 0.6363 which is most highly significant.

On the next page you can see how the data was growing during the last few months. You will find a tube around the forecast values. This is the range where the forecasts will possibly move. The safety probability of these confidence intervals is 95%.

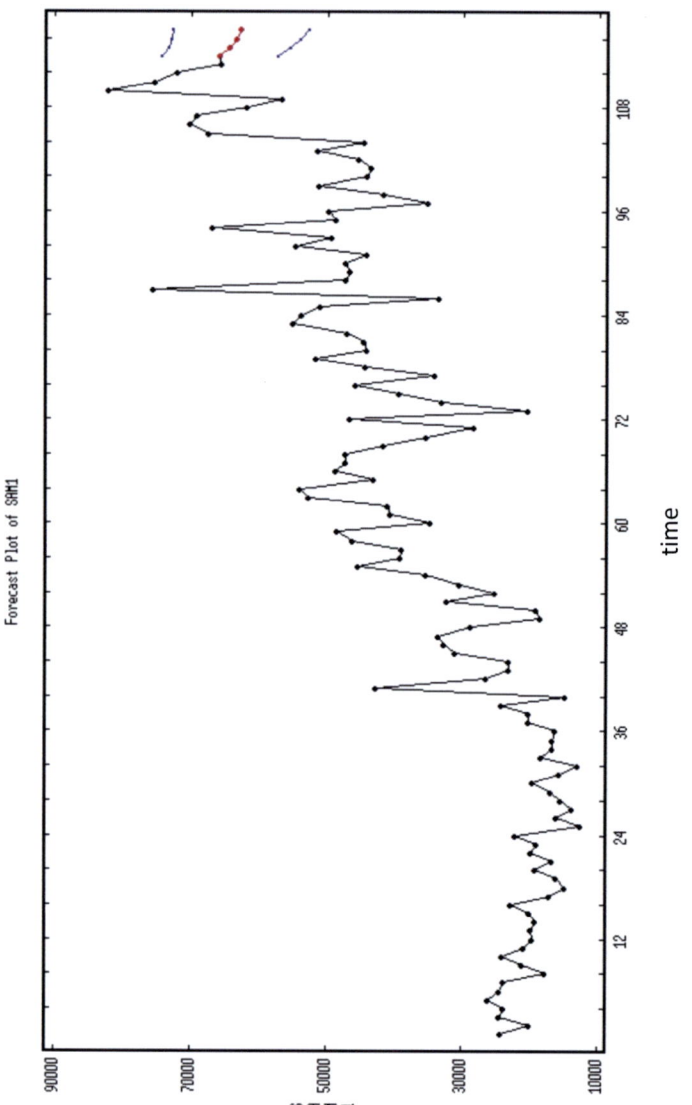

Fig. 48: Forecast Plot of Time Series SAM1

6.2   Analyzing and Forecasting Time Series SAM2

The last time series which is modeled and forecasted will be SAM2. SAM2 is data about an additional quantity to the amount of dangerous waste in Rheinland-Pfalz. The analysis and forecasting will be based on the recommended macro structure, which was introduced in section 3.5.3.1: *Structure of the Spreadsheet Task for the Time Series ROEHM.*

The data of time series are listed below. We saved this time series as ASCII-file in our working-directory under the name SAMHERD2.dat

SAM2

708717	730216	881323	612434	823488	777484	863486	920263	902474	871949
910865	786633	573128	624597	788039	759398	688136	656575	747635	672993
718776	699143	667994	1133447	455398	479855	606792	612942	449928	554815
631958	504358	553195	586408	633965	573566	456985	312502	479250	341501
359239	384153	405545	346501	389564	420554	389766	384434	231055	252932
333538	342438	303325	298243	311718	281582	387163	345750	344925	341091
303342	319034	338372	350651	400419					

Number of measurements n = 65.

Fig. 49: Time Series SAM2

Note: The data have to be read from left to right, line by line. You will find the data macro in the Appendix 4.

Graphical Depicting of the Time Series SAM2 with SCAGRAF

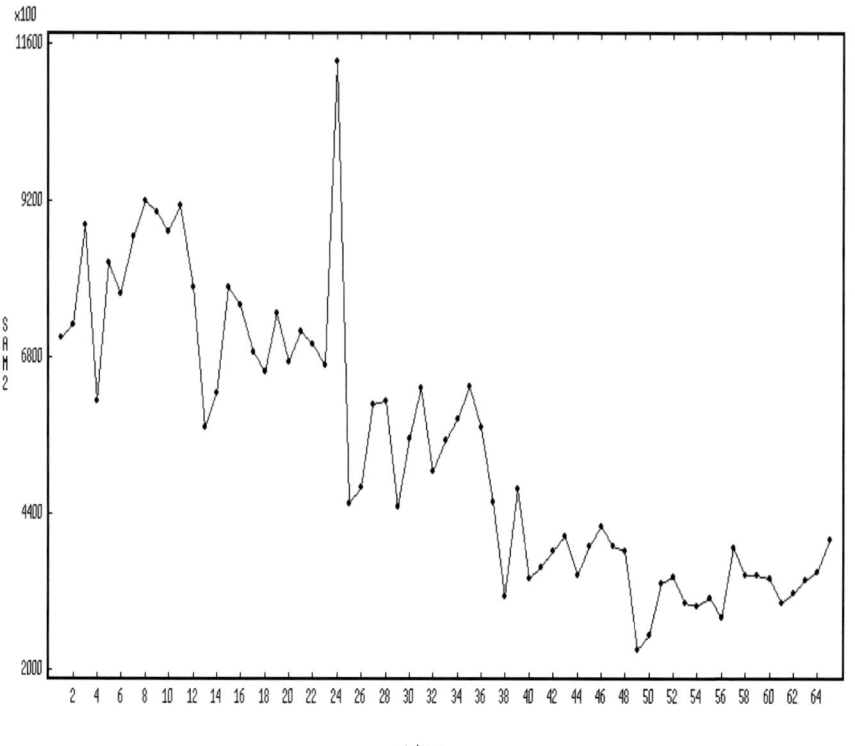

time

Fig. 50: Single Series Plot of Time Series SAM2

Note: The graph was made with SCAGRAF as explained in chap.4.

It seems that there is one outlier at t = 24 in the time series SAM2, but this should be taken care of in the macro procedure, with the OUTLIER and OESTIM command. Because of the non stationarity of the data, it seems that differencing is necessary. The ACF of the series SAM2 is depicted below.

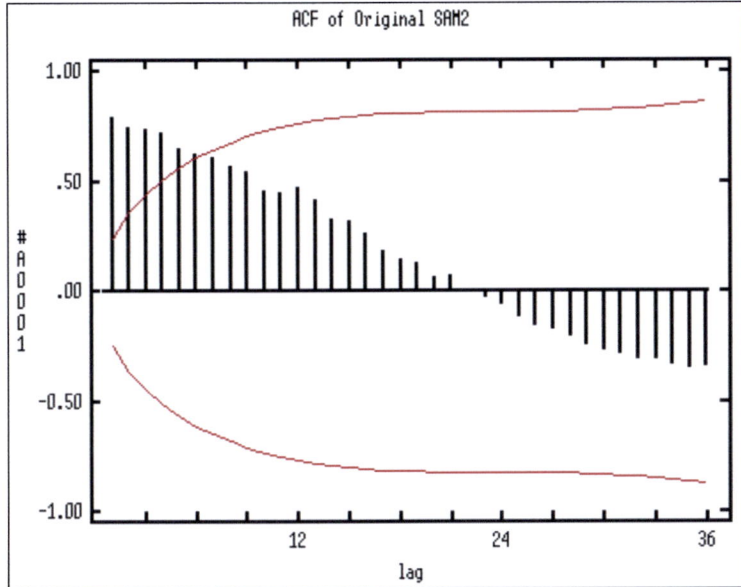

Fig. 51:ACF of the Time Series SAM2

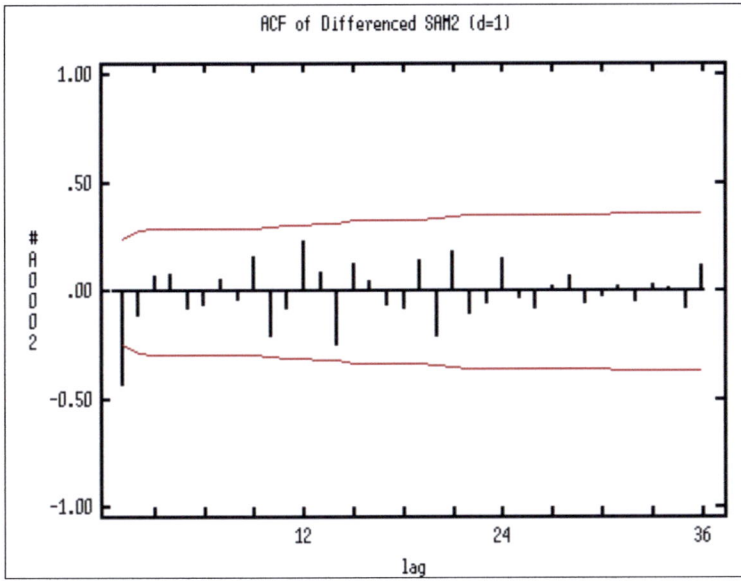

Fig. 52: ACF of the Differenced Time Series SAM2

The ACF shows no signs of seasonality in the series. The ACF of the differenced series lets one assume that there is a MA- part in the model of the series because it dies out. So it seems, that the model will be of the type ARIMA (0,1,1).

Forecasting the Time Series SAM2

The recommended macro for forecasting this time series is listed below. You will find the macro DATA and the file SAMHERD2.mad in the Appendix 4. In the case of the time series SAM2 we decided to employ the ln-transformation, because without this transformation the standard error was very big and therefore the model was not reliable anymore. We also tried to find a user defined model which might be better than automatic moelbuilding. Even after several trials the model from the automatic modeling had better standard errors. Therefore we decided just to include the automatic modeling in our macro procedure.

Note: After employing the ln-transformation the results of the forecast have to be transferred back with the exponential function according to the formula in chap.4: *Annotation: Calculation of the Confidence Interval.*

```
==ALLMACRO
   ASSIGN FILE 12. EXTERNAL 'SAMHERD2.MAD'.
   CALL DATA. FILE 12.
C TIME SERIES VARIALE SAM2 MAY BE REFERENCED BY POSITIONAL VARIABLE:
C &V_1.
   CALL ANALYSIS
   STOP
==ANALYSIS


C AUTOMATIC MODELING
&V_1=LN(&V_1)
IARIMA &V_1
C AUTOMATIC MODELING OF THE LN- TRANSFORMED DATA


ACF &V_1. MAXLAG IS 36
C DEPICTING OF THE ACF FOR THE TIME SERIES LN(SAM1)
```

```
OUTLIER UTSMODEL. TYPES ARE AO,IO,LS,TC
C OUTLIER DETECTION OF THE MODEL CREATED BY THE IARIMA COMMAND
OESTIM UTSMODEL. METHOD IS EXACT
C ESTIMATION OF THE MODEL PARAMETERS INCLUDING THE INFORMATION OF THE
C OUTLIERS
       NEW SERIES IN V1, V2, V3.
C SAVES THE INFORMATION DURING THE OUTLIER DETECTION PROCESS
C V1: NAME USED TO STORE RESIDUALS AFTER OUTLIER ADUSTMENT
C V2: NAME FOR THE ADJUSTED SERIES
C V3: NAME FOR THE INDICATOR VARIABLES WITH THE TYPE OF THE
C OUTLIER.
C V2 AND V2 WILL BE NEEDED FOR THE SCAGRAF SESSION TO PRINT
C THE OUTLIER PLOT

FORECAST UTSMODEL.NOFS ARE 4
C FORECAST FOR TIME SERIES MODEL FOUND BY IARIMA

OFORECAST UTSMODEL.NOFS ARE 4
C FORECAST OF THE LN-TRANSFORMED TIME SERIES SAM2 INCLUDING THE OUTLIER
C INFORMATION

 RETURN
```

You will find the complete output file of the macro procedure in the Appendix 4 .

The different AIC values are

AIC1=2 ln 0.188498 + ln(65)/65 = -3.27: Automatic modeling without outlier adjustment

AIC2=2 ln 0.120408 + ln(65)/65 = -4.16: Automatic modeling with outlier adjustment

AIC2 has the smaller value and therefore the results of the session with outlier adjustment will be included in our standard report for top managers.

Since we employed the ln-transformation we have to re-transform the forecasted values with the exponential function. The forecasts according to the output file in the Appendix 4 are:

```
TIME      FORECAST    STD. ERROR
  66      12.7821       0.1204
  67      12.7825       0.1257
  68      12.7827       0.1307
  69      12.7829       0.1356
```

Therefore the values of the forecast according to the formula in chap.4 *(Annotation: Calcula-tion of the Confidence Interval)* are

time 66 => $e^{12.7821} = 355791.441$

time 67 => $e^{12.7825} = 355933.786$

time 68 => $e^{12.7827} = 356004.980$

time 69 => $e^{12.7829} = 356076.188$.

With the safety probability S = 70% $\Rightarrow$ c(S) = 1.036 we generated the following confidence interval (CI) for time 66

CI = 12.7821 ± 1.036*0.1204 = 12.7821 ± 0.12473 =>

CI1 = 12.9068

CI2 = 12.6573.

Therefore the limits of the confidence interval for the original measurements are

CI1 = exp(12.9068) = 403043.590 (upper limit)

CI2 = exp(12.6573) = 314047.648 (lower limit).

The corresponding values for the other times are computed in the same way.

STANDARD REPORT FOR TOP MANAGERS FOR THE TIME SERIES SAM2

The expected values for the waste volume in the next 4 months are:

	Forecast	L.L. of the CI with 70% safety probability	U.L. of the CI with 70% safety probability
Month 01	355791.441	314047.648	403043.590
Month 02	355933.786	312450.082	405428.576
Month 03	356004.980	310907.276	407623.812
Month 04	356076 .188	309409.214	409781.758

CI: Confidence Interval of the Forecast

LL: Lower Limit of CI

UL: Upper Limit of CI.

7 Summary and Outlook

In the present publication we discussed a method to analyse and forecast time series. With this method it is possible to plan for the future. This is very important for logistics in the manner of the CHNL (KGVL®), because of this the complexity in the logistics field decreases. The principle of the logistics flow is easier to turn into reality. With forecast values it is easier to fulfill the logistics task, because two elements (the right amount and the right time) are known in the preliminary stage.

In chapter 3 and 4 the software solution of the SCA-System for forecasting regular time series is considered. Because of the user-friendly surface and the possibility to create macros, it is easier to analyse and forecast time series. In chapter 5 a recommendation is made, how to use this specific software tool. The following chapter 6 includes the forecast of two examples taken from the logistics field.

In the Appendix 5 the analysis of irregular time series is discussed shortly. A method which was developed by the NATIONAL SIENCE FOUNDATION (NFS) is mentioned, which discusses the forecast of those irregular time series.

During our work we realized, that it would be of great importance for several companies, if they would be able to analyse and forecast their data. A BMW dealer had made a very interesting comment. He told us, that BMW created a new software which will forecast the demand of spare parts. If a dealer who uses this software, orders to much, he can send the surplus back for free. Unfortunately the surface of the software was not user-friendly and the system was complicated to use. So the BMW dealer decided not to purchase it.

To mention is, that the BMW dealer is knowledgeable with the use of computers and also has extensive working knowledge in using the Internet and the products of MS-MICROSOFT OFFICE.

On the other side there was a company, which had a very negative attitude against statistics. "I know a lot about statistics and anyways the data are never correct, because the reality is different and so I have no interest in it." (Original quote of a staff manager). But this manager had no idea, how the software of SCA works. He never heard about the BOX-JENKINS-method, ARIMA-models and the BOOTSTRAPPING method.

Many good books about the time series and methods are mostly from the United States. Just a fraction of it is translated into the German language. Also good software solutions are from the United States or the United Kingdom, whereas the big German software producers, for example SAP, still use methods which are known by everybody.

But most of the managers of small and middle companies to whom we explained the aim of the present publication showed a large interest in software solutions to forecast their data. So we just want to mention that "not the big one eats the small one, instead the fast one eats the slow one".

Appendix

Appendix 1: Guide for Translating ARIMA/SARIMA-Models into SCA-Convention

<u>ARIMA-Models</u>

We have an ARIMA (p, d, q) model , which is

$$(1-B)^d Z_t = \mu + [\theta(B) / \varphi(B)] * a_t$$

in operator form, with

        p = degree of the non-seasonal AR- polynomial with parameters $\varphi_1, \varphi_2,..., \varphi_p$

        d = order of non-seasonal differencing

        q = degree of the non-seasonal MA- polynomial with parameters $\theta_1, \theta_2,..., \theta_q$.

If this model has the form ARIMA (1, 1, 1), it would be

$$(1-\varphi_1 *B)*(1-B)\ Z_t = C + (1-\theta_1 *B) * a_t$$
$$C = \mu * (1-\varphi_1).$$

In SCA-convention this model is

$$(1-PHI1*B)\ \text{name of TS } (1) = CONST + (1-THETA1*B)\ NOISE$$

      $\downarrow$ $\downarrow$ $\downarrow$

    p = 1   d = 1   q = 1

<u>SARIMA-Models</u>

We have a SARIMA (p, d, q) x (P, D, Q) model , which is

$$(1-B)^d (1-B^s)^D Z_t = \mu + [\theta(B)\Theta(B^s) / (\varphi(B)\, \Phi(B^s))] * a_t$$

in operator form, with

        p = degree of the non-seasonal AR- polynomial

        d = order of non-seasonal differencing

        q = degree of the non-seasonal MA- polynomial

P = order of the seasonal AR- part with parameters $\Phi_1, \Phi_{2,...}, \Phi_P$

D = order of seasonal differencing

Q = order of the seasonal MA- part with parameters $\Theta_1, \Theta_{2,...}, \Theta_Q$

s = length of the season.

If this model has the form SARIMA (1, 1, 1) x (1, 1, 1,) with s=12, it would be

$(1-\varphi_1 * B) \; (1-\Phi_1*B^{12})(1-B)(1- B^{12})*Z_t = C + (1-\theta_1 * B) \; (1-\Theta_1*B^{12})* a_t$

$C = \mu * (1-\varphi_1) (1-\Phi_1).$

In SCA- convention this model is

(1-PHI1*B)(1-PHI12*B**12)name of TS(1,12)=CONST+(1-THETA1*B)(1-THETA12*B**12) NOISE

$\downarrow \quad\quad \downarrow \quad\quad\quad\quad\quad\quad\quad \downarrow \quad \downarrow \quad\quad\quad\quad \downarrow \quad\quad\quad\quad\quad \downarrow$

p=1   P=1   d = 1 D=1, s = 12   q=1   Q=1

For the time series ROEHM with the model SARIMA(0,1,1)x(0,1,1) therefore is valid

$(1-B) (1-B^{12}) *Z_t = C + [(1-\theta B) * (1-\Theta B^{12})] * a_t$

with the SCA-convention and C=0:

ROEHM(1,12)=(1-THETA1*B)(1-THETA12*B**12)NOISE

Appendix 2: Output File for Time Series ROEHM

```
THE SCA STATISTICAL SYSTEM ( RELEASE VI.2B )
 SCA PRODUCT   IDENTIFICATION: GSA, UTS & QPI
 SCA PRODUCT   IDENTIFICATION: EXTENDED-UTS & EXPERT
 SCA SOFTWARE IDENTIFICATION: NOLLAU.H-REFA    ( 1041919 )
 SCA SOFTWARE RELEASE DATE:  3/ 1/2001
 SCA SOFTWARE RENEWAL DATE: 12/ 2/2006
 COPYRIGHT (C), SCIENTIFIC COMPUTING ASSOCIATES. ALL RIGHTS RESERVED
 SIZE OF WORKSPACE IS  4000000   SINGLE PRECISION WORDS
 DATE --  11/26/2001            TIME -- 18:02:39
 --
```

```
CALL ALLMACRO. FILE 'ROEHMTSA.MAC'
--

   ASSIGN FILE 12. EXTERNAL 'ROEHMTSA.MAD'.
--

C ROEHMTSA.MAD IS THE FILE WHICH CONTAINS THE DATA OF THE TIME
C SERIES
--

CALL DATA. FILE 12.
--

INPUT ROEHM. @
   PREC DOUB. @
   FORMAT FREE(1,132).

ROEHM    , A  155  BY   1 VARIABLE, IS STORED IN THE WORKSPACE; DOUBLE PRE-
CISION
--

RETURN
--

C READING THE DATA OF TIME SERIES ROEHM INTO THE SCA-SYSTEM
--

C THE VARIABLE ROEHM MAY BE REFERENCED BY POSITIONAL VARIABELE: &V_1
--

CALL ANALYSIS
--

C AUTOMATIC MODELING
--

IARIMA &V_1.SEASONALITY IS 12

THE FOLLOWING ANALYSIS IS BASED ON TIME SPAN   1  THRU  155
THE CRITICAL VALUE FOR SIGNIFICANCE TESTS OF ACF AND ESTIMATES IS 1.960

SUMMARY FOR UNIVARIATE TIME SERIES MODEL -- UTSMODEL
----------------------------------------------------------------------
VARIABLE   TYPE OF    ORIGINAL     DIFFERENCING
           VARIABLE   OR CENTERED
                                        12
ROEHM      RANDOM     ORIGINAL     (1-B  )
----------------------------------------------------------------------
PARAMETER  VARIABLE  NUM./ FACTOR  ORDER   CONS-      VALUE     STD    T
  LABEL      NAME    DENOM.                TRAINT             ERROR VALUE
```

```
1                CNST    1    0    NONE   4857.9930  264.4262  18.37
2        ROEHM   MA      1   12    NONE       .7999     .0881   9.08
3        ROEHM   D-AR    1   12    NONE       .3520     .1192   2.95
```

```
TOTAL NUMBER OF OBSERVATIONS . . . .       155
EFFECTIVE NUMBER OF OBSERVATIONS . .       131
RESIDUAL STANDARD ERROR. . . . . . .  0.623794E+04
--
C AUTOMATIC MODELING OF THE DATA
--
```

```
ACF &V_1. MAXLAG IS 36
```

```
NAME OF THE SERIES . . . . . . . . .          ROEHM
TIME PERIOD ANALYZED . . . . . . . . . 1  TO   155
MEAN OF THE (DIFFERENCED) SERIES . . .   67396.6094
STANDARD DEVIATION OF THE SERIES . . .   18816.6016
T-VALUE OF MEAN (AGAINST ZERO) . . . .      44.5926
```

```
AUTOCORRELATIONS


  1- 12     .86  .79  .78  .79  .76  .75  .73  .72  .68  .66  .73  .77
  ST.E.     .08  .13  .16  .18  .20  .22  .23  .25  .26  .27  .28  .29
  Q         116  216  312  413  508  599  687  774  850  923 1012 1113


 13- 24     .68  .62  .60  .60  .59  .56  .54  .54  .49  .47  .54  .55
  ST.E.     .31  .32  .32  .33  .34  .34  .35  .36  .36  .37  .37  .37
  Q        1192 1258 1321 1385 1446 1501 1554 1606 1650 1690 1743 1799


 25- 36     .49  .44  .41  .41  .39  .36  .35  .34  .28  .29  .32  .34
  ST.E.     .38  .38  .39  .39  .39  .39  .40  .40  .40  .40  .40  .41
  Q        1844 1881 1913 1945 1974 2000 2024 2047 2063 2080 2101 2125


         -1.0 -0.8 -0.6 -0.4 -0.2  0.0  0.2  0.4  0.6  0.8  1.0
         +----+----+----+----+----+----+----+----+----+----+
                                   I
  1  0.86                      +    IXXX+XXXXXXXXXXXXXXXX
  2  0.79                    +      IXXXXX+XXXXXXXXXXXXX
  3  0.78                 +         IXXXXXXX+XXXXXXXXXX
  4  0.79                +          IXXXXXXXX+XXXXXXXXX
  5  0.76               +           IXXXXXXXXX+XXXXXXXX
```

```
 6   0.75                    +              IXXXXXXXXXX+XXXXXXXX
 7   0.73                    +              IXXXXXXXXXX+XXXXXXX
 8   0.72                   +               IXXXXXXXXXXX+XXXXXX
 9   0.68                  +                IXXXXXXXXXXXX+XXXX
10   0.66                  +                IXXXXXXXXXXXX+XXX
11   0.73                 +                 IXXXXXXXXXXXXX+XXXX
12   0.77                 +                 IXXXXXXXXXXXXX+XXXXX
13   0.68                +                  IXXXXXXXXXXXXXX+XX
14   0.62               +                   IXXXXXXXXXXXXXXXX
15   0.60               +                   IXXXXXXXXXXXXXXXX+
16   0.60               +                   IXXXXXXXXXXXXXXXX+
17   0.59              +                    IXXXXXXXXXXXXXXXX +
18   0.56              +                    IXXXXXXXXXXXXXX  +
19   0.54              +                    IXXXXXXXXXXXXXX  +
20   0.54              +                    IXXXXXXXXXXXXX   +
21   0.49             +                     IXXXXXXXXXXXX       +
22   0.47             +                     IXXXXXXXXXXXX       +
23   0.54             +                     IXXXXXXXXXXXXX      +
24   0.55             +                     IXXXXXXXXXXXXXX     +
25   0.49            +                      IXXXXXXXXXXXX         +
26   0.44            +                      IXXXXXXXXXXX          +
27   0.41            +                      IXXXXXXXXXX           +
28   0.41            +                      IXXXXXXXXXX           +
29   0.39            +                      IXXXXXXXXXX           +
30   0.36            +                      IXXXXXXXXX            +
31   0.35            +                      IXXXXXXXXX            +
32   0.34           +                       IXXXXXXXXX             +
33   0.28           +                       IXXXXXXX               +
34   0.29           +                       IXXXXXXX               +
35   0.32           +                       IXXXXXXXX              +
36   0.34           +                       IXXXXXXXXX             +
--
C DEPICTING OF THE ACF FOR THE TIME SERIES ROEHM
--


OUTLIER UTSMODEL. TYPES ARE AO,IO,LS

INITIAL RESIDUAL STANDARD ERROR =   6034.6

TIME    ESTIMATE    T-VALUE    TYPE
```

```
   102  20185.57     3.49      IO
```

ADJUSTED RESIDUAL STANDARD ERROR =   5769.1

--

C OUTLIER DETECTION OF THE MODEL CREATED BY THE IARIMA COMMAND

--

OESTIM UTSMODEL. METHOD IS EXACT

THE FOLLOWING ANALYSIS IS BASED ON TIME SPAN   1 THRU 155

SUMMARY FOR UNIVARIATE TIME SERIES MODEL -- UTSMODEL

VARIABLE   TYPE OF   ORIGINAL   DIFFERENCING
           VARIABLE   OR CENTERED
                                      12
ROEHM   RANDOM   ORIGINAL   (1-B)

PARAMETER LABEL	VARIABLE NAME	NUM./ DENOM.	FACTOR	ORDER	CONS- TRAINT	VALUE	STD ERROR	T VALUE
1		CNST	1	0	NONE	4812.5895	254.8452	18.88
2	ROEHM	MA	1	12	NONE	.8618	.0598	14.40
3	ROEHM	D-AR	1	12	NONE	.3788	.0946	4.00

SUMMARY OF OUTLIER DETECTION AND ADJUSTMENT

TIME   ESTIMATE   T-VALUE   TYPE
----
 102   19960.961   3.34   IO

TOTAL NUMBER OF OBSERVATIONS.   155
EFFECTIVE NUMBER OF OBSERVATIONS.   131
RESIDUAL STANDARD ERROR (WITHOUT OUTLIER ADJUSTMENT). . 0.621820E+04
RESIDUAL STANDARD ERROR (WITH OUTLIER ADJUSTMENT) . . . 0.596801E+04

--

C ESTIMATION OF THE MODELPARAMETERS WITHOUT OUTLIER ADJUSTMENT

--

     NEW SERIES IN V1, V2, V3.

```
--

C SAVES THE INFORMATION DURING THE OUTLIER DETECTION PROCESS

--

C V1: NAME USED TO STORE RESIDUALS AFTER OUTLIER ADJUSTMENT

--

C V2: NAME FOR THE ADJUSTED SERIES

--

C V3: NAME FOR THE INDICATOR VARIABLES WITH THE TYPE OF THE

--

C OUTLIER.

--

C V2 AND V3 WILL BE NEEDED FOR THE SCAGRAF SESSION TO PRINT

--

C THE OUTLIER PLOT

--


FORECAST UTSMODEL.NOFS ARE 4
----------------------------------
   4 FORECASTS, BEGINNING AT  155
----------------------------------
   TIME    FORECAST    STD. ERROR   ACTUAL IF KNOWN
   156   96195.2080    5968.0093
   157   79823.6294    5968.0093
   158   91516.5995    5968.0093
   159  101983.0783    5968.0093
--


C FORECAST FOR TIME SERIES FOUND BY THE IARIMA COMMAND

--


OFORECAST UTSMODEL.NOFS ARE 4

RESIDUAL  STANDARD  ERROR  (USES  DATA  UP  TO  THE  FIRST  FORECAST  ORIGIN)=
5968.0

   TIME    ESTIMATE    T-VALUE     TYPE
   102   19960.961      3.34       IO


----------------------------------
   4 FORECASTS, BEGINNING AT  155
```

```
----------------------------------
TIME    FORECAST   STD. ERROR   ACTUAL IF KNOWN
 156   96195.2080   5968.0093
 157   79823.6294   5968.0093
 158   91516.5995   5968.0093
 159  101983.0783   5968.0093
--

C FORECAST INCLUDING THE OUTLIER INFORMATION
--

C USER DEFINED MODELING
--

TSMODEL NAME IS MODELROE. MODEL IS &V_1(1,12)=(1-THETA1*B)@
(1-THETA12*B**12)NOISE
--

C EXPECTED MODEL TYPE BASED ON ANALYSES OF GRAPHIC DEPICTING
C AND FORMER SCA SESSIONS
--

ESTIM MODELROE. METHOD IS EXACT. HOLD RESIDUALS(RESR)


THE FOLLOWING ANALYSIS IS BASED ON TIME SPAN   1   THRU   155


NONLINEAR ESTIMATION TERMINATED DUE TO:
RELATIVE CHANGE IN THE STANDARD ERROR LESS THAN 0.1000D-02

SUMMARY FOR UNIVARIATE TIME SERIES MODEL -- MODELROE
-----------------------------------------------------------------------
VARIABLE   TYPE OF    ORIGINAL      DIFFERENCING
           VARIABLE   OR CENTERED
                                      1       12
ROEHM     RANDOM     ORIGINAL     (1-B  ) (1-B  )
-----------------------------------------------------------------------
PARAMETER  VARIABLE  NUM./  FACTOR  ORDER  CONS-    VALUE    STD     T
  LABEL     NAME     DENOM.                TRAINT           ERROR  VALUE
  1 THETA1   ROEHM    MA      1      1     NONE     .8544   .0432  19.80
  2 THETA12  ROEHM    MA      2      12    NONE     .5514   .0722   7.63


EFFECTIVE NUMBER OF OBSERVATIONS . .          142
R-SQUARE . . . . . . . . . . . . .          0.894
RESIDUAL STANDARD ERROR. . . . . .  0.612863E+04
--

C MODEL ESTIMATING
--
```

```
ACF RESR. MAXLAG IS 36
NAME OF THE SERIES . . . . . . . . . .          RESR
TIME PERIOD ANALYZED . . . . . . . . . 14  TO    155
MEAN OF THE (DIFFERENCED) SERIES . . .     232.1988
STANDARD DEVIATION OF THE SERIES . . .    6109.1274
T-VALUE OF MEAN (AGAINST ZERO) . . . .       0.4529


AUTOCORRELATIONS
  1- 12     -.13  .01  .03 -.03  .05  .16 -.06  .11  .02 -.15  .06  .08
  ST.E.      .08  .09  .09  .09  .09  .09  .09  .09  .09  .09  .09  .09
  Q          2.5  2.5  2.6  2.7  3.1  6.8  7.5  9.3  9.3 12.6 13.1 14.1


 13- 24     -.09  .07 -.09 -.10  .07 -.07 -.06  .08 -.00 -.23  .19 -.26
  ST.E.      .09  .09  .09  .09  .09  .09  .09  .09  .10  .10  .10  .10
  Q         15.3 16.2 17.5 19.1 19.9 20.8 21.5 22.6 22.6 31.6 37.6 48.9


 25- 36      .01  .13 -.08  .05 -.04 -.13  .03  .03 -.15  .05 -.00 -.06
  ST.E.      .11  .11  .11  .11  .11  .11  .11  .11  .11  .11  .11  .11
  Q         48.9 51.8 53.0 53.4 53.7 56.6 56.8 56.9 60.9 61.4 61.4 62.1


           -1.0 -0.8 -0.6 -0.4 -0.2  0.0  0.2  0.4  0.6  0.8  1.0
            +----+----+----+----+----+----+----+----+----+----+
                                     I
  1 -0.13                          +XXXI   +
  2  0.01                          +   I   +
  3  0.03                          +   IX  +
  4 -0.03                          +   XI  +
  5  0.05                          +   IX  +
  6  0.16                          +   IXXXX
  7 -0.06                          + XXI   +
  8  0.11                          +   IXXX+
  9  0.02                          +   I   +
 10 -0.15                          XXXXI   +
 11  0.06                          +   IX  +
 12  0.08                          +   IXX +
 13 -0.09                          + XXI   +
 14  0.07                          +   IXX +
 15 -0.09                          + XXI   +
 16 -0.10                          + XXI   +
 17  0.07                          +   IXX +
```

```
18  -0.07                          +  XXI   +
19  -0.06                          +  XXI   +
20   0.08                          +   IXX  +
21   0.00                          +   I    +
22  -0.23                      X+XXXXI      +
23   0.19                          +   IXXXXX
24  -0.26                      X+XXXXI      +
25   0.01                          +   I    +
26   0.13                          +   IXXX +
27  -0.08                          +  XXI   +
28   0.05                          +   IX   +
29  -0.04                          +   XI   +
30  -0.13                          + XXXI   +
31   0.03                          +   IX   +
32   0.03                          +   IX   +
33  -0.15                          +XXXXI   +
34   0.05                          +   IX   +
35   0.00                          +   I    +
36  -0.06                          +  XXI   +
--
```

C DEPICTING OF RESIDUALS OF TIME SERIES ROEHM WITH THE

`--`

C EXPECTED MODEL

`--`

OESTIM MODELROE. METHOD IS EXACT

THE FOLLOWING ANALYSIS IS BASED ON TIME SPAN   1   THRU   155

SUMMARY FOR UNIVARIATE TIME SERIES MODEL -- MODELROE

```
----------------------------------------------------------------------
VARIABLE    TYPE OF    ORIGINAL    DIFFERENCING
            VARIABLE   OR CENTERED

                                      1      12
ROEHM       RANDOM     ORIGINAL    (1-B  ) (1-B  )
----------------------------------------------------------------------
```

PARAMETER LABEL	VARIABLE NAME	NUM./ DENOM.	FACTOR	ORDER	CONS- TRAINT	VALUE	STD ERROR	T VALUE
1 THETA1	ROEHM	MA	1	1	NONE	.8283	.0488	16.97
2 THETA12	ROEHM	MA	2	12	NONE	.5125	.0773	6.63

SUMMARY OF OUTLIER DETECTION AND ADJUSTMENT

```
-------------------------------------
TIME     ESTIMATE    T-VALUE    TYPE
-------------------------------------

102   18313.927      3.72       AO
125   16790.133      3.39       AO
-------------------------------------
```

```
TOTAL NUMBER OF OBSERVATIONS. . . . . . . . . . . . .        155
EFFECTIVE NUMBER OF OBSERVATIONS. . . . . . . . . . .        142
RESIDUAL STANDARD ERROR (WITHOUT OUTLIER ADJUSTMENT). .  0.645085E+04
RESIDUAL STANDARD ERROR (WITH OUTLIER ADJUSTMENT) . . .  0.592965E+04
--

C PARAMETER ESTIMATION AND ADUSTING THE USER DEFINED MODEL TO OUTLIERS
--

NEW SERIES IN X1, X2, X3.
--

C SAVES THE INFORMATION DURING THE OUTLIER DETECTION PROCESS
--

C X1: NAME USED TO STORE RESIDUALS AFTER OUTLIER ADJUSTMENT
--

C X2: NAME FOR THE ADJUSTED SERIES
--

C X3: NAME FOR THE INDICATOR VARIABLES WITH THE TYPE OF THE
--

C OUTLIER.
--

C X2 AND X3 WILL BE NEEDED FOR THE SCAGRAF SESSION TO PRINT
--

C THE OUTLIER PLOT
--

FORECAST MODELROE.NOFS ARE 4
---------------------------------
   4 FORECASTS, BEGINNING AT   155
---------------------------------
TIME     FORECAST    STD. ERROR    ACTUAL IF KNOWN
 156   95587.2483    5929.6538
 157   75929.4256    6016.4412
 158   89939.9024    6101.9943
 159  100657.4668    6186.3644
--
```

```
C FORECAST FOR THE USER DEFINED MODEL

--

OFORECAST MODELROE.NOFS ARE 4

RESIDUAL  STANDARD  ERROR  (USES  DATA  UP  TO  THE  FIRST  FORECAST  ORIGIN)=
5789.1


   TIME     ESTIMATE   T-VALUE     TYPE
    102   18240.024      3.80       AO
    125   16562.615      3.42       AO
    153   15209.161      2.69       AO
----------------------------------
   4 FORECASTS, BEGINNING AT   155
----------------------------------
   TIME     FORECAST   STD. ERROR    ACTUAL IF KNOWN
    156   94161.9251   5929.6538
    157   74504.1001   6016.4412
    158   88514.5770   6101.9943
    159   99232.1413   6186.3644
  --
C FORECAST FOR OUTLIER ADJUSTED USER DEFINED MODEL

--

   RETURN

--

   STOP


THE CURRENT SCA SESSION IS TERMINATED.
THE SIZE OF THE WORKSPACE USED IS  47221 WORDS.
```

Appendix 3: Data Macro and Output File for Time Series SAM1
The Data Macro saved in File SAMHERD.MAD

```
==DATA
INPUT serie1. @
  PREC DOUB. @
  FORMAT FREE(1,132).
24293
20137
24527
23883
26053
```

```
24422
23717
17756
21063
24020
20868
19616
19869
19227
20179
22728
17252
14800
16108
19236
16673
19812
19095
22189
12506
16136
13876
15571
16970
19673
15775
12918
18313
16810
16689
16322
20221
20379
24244
14883
42820
26542
23101
23120
31006
```

32779

33618

28822

18672

19177

32345

25367

30566

35542

45467

39213

38986

46243

48578

34923

40678

41027

52723

53860

43044

48852

47252

47247

41747

35492

28353

46582

20467

33218

39375

45857

34169

44386

51620

44168

44523

46992

54935

53734

51140

```
33607
75473
47366
46623
47390
44261
54624
49274
66861
48833
49886
35300
41712
51252
44115
43592
45510
51442
44571
67351
70141
69070
61841
56749
82140
75391
72115
65619
End of Data
-- This data macro was created on: 22.06.2001 14:49:02
-- Number of data rows written: 113
Return
```

The Output file for Time Series SAM1 (Saved in File SAM1.OTP)

```
THE SCA STATISTICAL SYSTEM ( RELEASE VI.2B )
 SCA PRODUCT  IDENTIFICATION: GSA, UTS & QPI
 SCA PRODUCT  IDENTIFICATION: EXTENDED-UTS & EXPERT
 SCA SOFTWARE IDENTIFICATION: NOLLAU.H-REFA    ( 1041919 )
 SCA SOFTWARE RELEASE DATE:  3/ 1/2001
 SCA SOFTWARE RENEWAL DATE: 12/ 2/2006
 COPYRIGHT (C), SCIENTIFIC COMPUTING ASSOCIATES. ALL RIGHTS RESERVED
```

```
SIZE OF WORKSPACE IS  4000000   SINGLE PRECISION WORDS
DATE -- 11/26/2001              TIME -- 18:12:12
--
CALL ALLMACRO. FILE 'SAMHERD.MAC'
--
  ASSIGN FILE 12. EXTERNAL 'SAMHERD.MAD'.
--
  CALL DATA. FILE 12.
--
INPUT SERIE1. @
  PREC DOUB. @
  FORMAT FREE(1,132).
SERIE1  , A 113  BY   1 VARIABLE, IS STORED IN THE WORKSPACE; DOUBLE PRECISION
--
RETURN
--
C TIME SERIES VARIABLE SERIE1 MAY BE REFERENCED BY POSITIONAL
C VARIABLE: &V_1.
--
  CALL ANALYSIS
--
C AUTOMATIC MODELING
--
IARIMA &V_1.SEASONALITY IS 12

THE FOLLOWING ANALYSIS IS BASED ON TIME SPAN   1  THRU  113
THE CRITICAL VALUE FOR SIGNIFICANCE TESTS OF ACF AND ESTIMATES IS 1.960
SUMMARY FOR UNIVARIATE TIME SERIES MODEL -- UTSMODEL
----------------------------------------------------------------------
VARIABLE   TYPE OF    ORIGINAL     DIFFERENCING
           VARIABLE   OR CENTERED
                                        1
  SERIE1   RANDOM     ORIGINAL     (1-B )
----------------------------------------------------------------------
  PARAMETER   VARIABLE   NUM./  FACTOR  ORDER   CONS-    VALUE    STD     T
    LABEL       NAME     DENOM.                 TRAINT            ERROR   VALUE
    1          SERIE1    MA       1       1     NONE     .6284   .0735   8.55

TOTAL NUMBER OF OBSERVATIONS . . . .        113
EFFECTIVE NUMBER OF OBSERVATIONS . .        112
RESIDUAL STANDARD ERROR. . . . . .  0.787254E+04
--
C AUTOMATIC MODELING OF THE DATA
--
ACF &V_1. MAXLAG IS 36
```

```
NAME OF THE SERIES . . . . . . . . .        SERIE1
TIME PERIOD ANALYZED . . . . . . . .   1  TO   113
MEAN OF THE (DIFFERENCED) SERIES . . .  35922.5313
STANDARD DEVIATION OF THE SERIES . . .  16671.3613
T-VALUE OF MEAN (AGAINST ZERO) . . . .     22.9053
```

AUTOCORRELATIONS

```
 1- 12   .82  .79  .74  .73  .69  .64  .61  .56  .53  .49  .52  .51
ST.E.     .09  .14  .18  .20  .23  .24  .26  .27  .28  .29  .30  .30
   Q      78.9 153  218  281  338  387  434  472  507  537  571  604


13- 24   .49  .44  .46  .46  .46  .47  .42  .41  .39  .36  .38  .36
ST.E.     .31  .32  .32  .33  .34  .34  .35  .35  .36  .36  .36  .37
   Q      635  661  689  717  746  776  801  824  845  864  885  904


25- 36   .33  .30  .25  .24  .21  .16  .15  .09  .07  .04  .03  .01
ST.E.     .37  .37  .37  .38  .38  .38  .38  .38  .38  .38  .38  .38
   Q      919  933  943  952  958  962  966  967  968  968  969  969
```

```
         -1.0 -0.8 -0.6 -0.4 -0.2  0.0  0.2  0.4  0.6  0.8  1.0
         +----+----+----+----+----+----+----+----+----+----+
                                    I
  1  0.82                      +    IXXXX+XXXXXXXXXXXXXXXX
  2  0.79                     +     IXXXXXX+XXXXXXXXXXXXX
  3  0.74                   +       IXXXXXXXX+XXXXXXXXXX
  4  0.73                  +        IXXXXXXXXX+XXXXXXXX
  5  0.69                 +         IXXXXXXXXXX+XXXXXX
  6  0.64                +          IXXXXXXXXXXX+XXXX
  7  0.61               +           IXXXXXXXXXXXX+XX
  8  0.56               +           IXXXXXXXXXXXX+X
  9  0.53              +            IXXXXXXXXXXXXX+
 10  0.49              +            IXXXXXXXXXXXX +
 11  0.52             +             IXXXXXXXXXXXXX +
 12  0.51             +             IXXXXXXXXXXXXX +
 13  0.49             +             IXXXXXXXXXXXX  +
 14  0.44            +              IXXXXXXXXXXX      +
 15  0.46            +              IXXXXXXXXXXX     +
 16  0.46            +              IXXXXXXXXXXX     +
 17  0.46            +              IXXXXXXXXXXX     +
 18  0.47           +               IXXXXXXXXXXX      +
 19  0.42           +               IXXXXXXXXXXX     +
 20  0.41           +               IXXXXXXXXXX      +
 21  0.39           +               IXXXXXXXXXX      +
```

```
22   0.36          +                IXXXXXXXXX       +
23   0.38          +                IXXXXXXXXX       +
24   0.36          +                IXXXXXXXXX       +
25   0.33          +                IXXXXXXXX        +
26   0.30          +                IXXXXXXX         +
27   0.25          +                IXXXXXX          +
28   0.24          +                IXXXXXX          +
29   0.21          +                IXXXXX           +
30   0.16          +                IXXXX            +
31   0.15        +                  IXXXX              +
32   0.09        +                  IXX                +
33   0.07        +                  IXX                +
34   0.04        +                  IX                 +
35   0.03        +                  IX                 +
36   0.01        +                  I                  +
--

C DEPICTING OF THE ACF FOR THE TIME SERIES SAM1
--
OUTLIER UTSMODEL. TYPES ARE AO,IO,LS,TC
INITIAL RESIDUAL STANDARD ERROR =   7794.0

   TIME    ESTIMATE    T-VALUE    TYPE
     87    29347.37       4.46    AO
    105    21518.12       3.98    LS
     41    24197.81       3.79    IO
     73   -19071.55      -3.48    AO
     94    19185.71       3.65    AO
    110    19448.26       3.59    IO


ADJUSTED RESIDUAL STANDARD ERROR =   5385.8
--


C OUTLIER DETECTION OF THE MODEL CREATED BY THE IARIMA COMMAND
--
OESTIM UTSMODEL. METHOD IS EXACT
THE FOLLOWING ANALYSIS IS BASED ON TIME SPAN   1   THRU   113


SUMMARY FOR UNIVARIATE TIME SERIES MODEL -- UTSMODEL


------------------------------------------------------------------------
VARIABLE   TYPE OF    ORIGINAL    DIFFERENCING
           VARIABLE   OR CENTERED
                                        1
SERIE1     RANDOM     ORIGINAL    (1-B )
------------------------------------------------------------------------
```

PARAMETER LABEL	VARIABLE NAME	NUM./ DENOM.	FACTOR	ORDER	CONS-TRAINT	VALUE	STD ERROR	T VALUE
1	SERIE1	MA	1	1	NONE	.6363	.0773	8.23

SUMMARY OF OUTLIER DETECTION AND ADJUSTMENT

```
------------------------------------
TIME    ESTIMATE    T-VALUE    TYPE
------------------------------------
  41   24193.602      5.61     IO
  49  -15426.919     -4.07     TC
  63   12000.624      3.16     TC
  71  -15240.742     -3.82     AO
  73  -21780.416     -5.61     TC
  86  -17814.837     -4.13     IO
  87   29235.654      7.49     AO
  94   17504.512      4.44     AO
  97  -14077.230     -3.23     IO
 105   20154.498      6.03     LS
 110   19662.189      4.83     TC
------------------------------------
```

MAXIMUM NUMBER OF OUTLIERS IS REACHED

** THE OUTLIER(S) AFTER TIME PERIOD 108   OCCURS WITHIN THE
   LAST FIVE OBSERVATIONS OF THE SERIES. THE IDENTIFIED TYPE
   ANS THE ESTIMATE OF THE OUTLIER(S) MAY NOT BE RELIABLE

TOTAL NUMBER OF OBSERVATIONS.   113
EFFECTIVE NUMBER OF OBSERVATIONS.   112
RESIDUAL STANDARD ERROR (WITHOUT OUTLIER ADJUSTMENT). . 0.790670E+04
RESIDUAL STANDARD ERROR (WITH OUTLIER ADJUSTMENT) . . . 0.431469E+04
--
C ESTIMATION OF THE MODEL INCLUDING THE
C INFORMATION OF THE OUTLIERS
--
     NEW SERIES IN V1, V2, V3.
--
C SAVES THE INFORMATION DURING THE OUTLIER DETECTION PROCESS
--
C V1: NAME USED TO STORE RESIDUALS AFTER OUTLIER ADUSTMENT
--
C V2: NAME FOR THE ADJUSTED SERIES
--

C V3: NAME FOR THE INDICATOR VARIABLE WITH THE TYPE OF THE

--

C OUTLIER.

--

C V2 AND V3 WILL BE NEEDED FOR THE SCAGRAF SESSION TO PRINT

--

C THE OUTLIER PLOT

--

FORECAST UTSMODEL.NOFS ARE 4

----

    4 FORECASTS, BEGINNING AT 113

    TIME   FORECAST   STD. ERROR   ACTUAL IF KNOWN

    114   69278.7105   4314.6927
    115   69278.7105   4591.2345
    116   69278.7105   4852.0403
    117   69278.7105   5099.5251

--

C FORECAST FOR TIME SERIES MODEL FOUND BY IARIMA

--

OFORECAST UTSMODEL.NOFS ARE 4

RESIDUAL STANDARD ERROR (USES DATA UP TO THE FIRST FORECAST ORIGIN)= 4314.7

TIME	ESTIMATE	T-VALUE	TYPE
41	24193.602	5.61	IO
49	-15426.919	-4.07	TC
63	12000.624	3.16	TC
71	-15240.742	-3.82	AO
73	-21780.416	-5.61	TC
86	-17814.837	-4.13	IO
87	29235.654	7.49	AO
94	17504.512	4.44	AO
97	-14077.230	-3.23	IO
105	20154.498	6.03	LS
110	19662.189	4.83	TC

MAXIMUM NUMBER OF OUTLIERS IS REACHED

----

    4 FORECASTS, BEGINNING AT 113

```
TIME     FORECAST    STD. ERROR    ACTUAL IF KNOWN

 114   65791.2644    4314.6927
 115   64374.9999    4591.2345
 116   63383.6147    4852.0403
 117   62689.6451    5099.5251
--

C FORECAST FIGURES INCLUDING THE OUTLIER INFORMATION
--


C USER DEFINED MODELING
--

TSMODEL NAME IS MODELSAM. MODEL IS &V_1(1)=(1-THETA1*B)NOISE
--

C EXPECTED MODEL TYPE BASED ON ANALYSES OF GRAPHIC DEPICTING
--

ESTIM MODELSAM. METHOD IS EXACT. HOLD RESIDUALS(RESR)

THE FOLLOWING ANALYSIS IS BASED ON TIME SPAN   1   THRU  113

NONLINEAR ESTIMATION TERMINATED DUE TO:
RELATIVE CHANGE IN THE STANDARD ERROR LESS THAN 0.1000D-02

SUMMARY FOR UNIVARIATE TIME SERIES MODEL -- MODELSAM
----------------------------------------------------------------------
VARIABLE    TYPE OF     ORIGINAL      DIFFERENCING
            VARIABLE    OR CENTERED
                                          1
 SERIE1     RANDOM      ORIGINAL      (1-B  )
----------------------------------------------------------------------
PARAMETER   VARIABLE   NUM./  FACTOR  ORDER   CONS-    VALUE    STD    T
  LABEL      NAME      DENOM.                  TRAINT           ERROR VALUE


 1  THETA1   SERIE1     MA      1       1     NONE     .6240   .0740  8.43


EFFECTIVE NUMBER OF OBSERVATIONS . .         112
R-SQUARE . . . . . . . . . . . . .          0.777
RESIDUAL STANDARD ERROR. . . . . .    0.787168E+04
--

C MODEL CREATING
--


ACF RESR. MAXLAG IS 36
```

```
NAME OF THE SERIES . . . . . . . . . .           RESR
TIME PERIOD ANALYZED . . . . . . . . . 2  TO    113
MEAN OF THE (DIFFERENCED) SERIES . . .    1084.6541
STANDARD DEVIATION OF THE SERIES . . .    7795.9658
T-VALUE OF MEAN (AGAINST ZERO) . . . .       1.4724
```

AUTOCORRELATIONS

```
 1- 12    -.06 -.00 -.07  .15  .05 -.09  .03 -.09 -.01 -.24  .11  .07
 ST.E.     .09  .09  .09  .10  .10  .10  .10  .10  .10  .10  .10  .11
 Q         .5   .5  1.0  3.7  4.0  5.0  5.1  6.1  6.1 13.4 15.0 15.6


13- 24     .01 -.26 -.01 -.00 -.01  .11 -.10  .04 -.01 -.04  .13  .11
 ST.E.     .11  .11  .11  .11  .11  .11  .11  .11  .11  .11  .11  .11
 Q        15.6 24.2 24.3 24.3 24.3 25.8 27.1 27.4 27.4 27.6 30.0 31.7


25- 36     .02 -.02  .06  .05  .00 -.05  .12 -.12 -.08 -.12 -.02 -.01
 ST.E.     .12  .12  .12  .12  .12  .12  .12  .12  .12  .12  .12  .12
 Q        31.8 31.8 32.3 32.6 32.7 33.1 35.3 37.4 38.4 40.8 40.8 40.9
```

```
             -1.0 -0.8 -0.6 -0.4 -0.2  0.0  0.2  0.4  0.6  0.8  1.0
             +----+----+----+----+----+----+----+----+----+----+
                                       I
   1  -0.06                       +   XXI   +
   2   0.00                       +    I    +
   3  -0.07                       +   XXI   +
   4   0.15                       +    IXXXX+
   5   0.05                       +    IX   +
   6  -0.09                       +   XXI   +
   7   0.03                       +    IX   +
   8  -0.09                       +   XXI   +
   9  -0.01                       +    I    +
  10  -0.24                   X+XXXXI   +
  11   0.11                       +    IXXX +
  12   0.07                       +    IXX  +
  13   0.01                       +    I    +
  14  -0.26                   X+XXXXI   +
  15  -0.01                       +    I    +
  16   0.00                       +    I    +
  17  -0.01                       +    I    +
  18   0.11                       +    IXXX +
  19  -0.10                       +  XXI    +
  20   0.04                   +        IX   +
  21  -0.01                   +        I    +
```

```
22  -0.04                    +    XI    +
23   0.13                    +    IXXX  +
24   0.11                    +    IXXX  +
25   0.02                    +    IX    +
26  -0.02                    +    I     +
27   0.06                    +    IX    +
28   0.05                    +    IX    +
29   0.00                    +    I     +
30  -0.05                    +    XI    +
31   0.12                    +    IXXX  +
32  -0.12                    +  XXXI    +
33  -0.08                    +   XXI    +
34  -0.12                    +  XXXI    +
35  -0.02                    +    I     +
36  -0.01                    +    I     +
--

C DEPICTING OF RESIDUALS OF TIME SERIES SAM1 WITH THE
--
C EXPECTED MODEL
--


OESTIM MODELSAM. METHOD IS EXACT
THE FOLLOWING ANALYSIS IS BASED ON TIME SPAN   1  THRU  113


SUMMARY FOR UNIVARIATE TIME SERIES MODEL -- MODELSAM
-----------------------------------------------------------------------
VARIABLE   TYPE OF    ORIGINAL     DIFFERENCING
           VARIABLE   OR CENTERED
                                        1
SERIE1    RANDOM     ORIGINAL     (1-B  )
-----------------------------------------------------------------------


PARAMETER   VARIABLE  NUM./ FACTOR  ORDER   CONS-     VALUE    STD     T
  LABEL      NAME    DENOM.                 TRAINT          ERROR  VALUE
  1 THETA1   SERIE1    MA      1      1     NONE     .6363   .0773  8.23


SUMMARY OF OUTLIER DETECTION AND ADJUSTMENT
------------------------------------
TIME    ESTIMATE   T-VALUE    TYPE
------------------------------------
  41   24193.603    5.61     IO
  49  -15426.918   -4.07     TC
  63   12000.625    3.16     TC
  71  -15240.739   -3.82     AO
  73  -21780.418   -5.61     TC
```

```
 86 -17814.840      -4.13      IO
 87  29235.658       7.49      AO
 94  17504.511       4.44      AO
 97 -14077.231      -3.23      IO
105  20154.501       6.03      LS
110  19662.195       4.83      TC
------------------------------------
```

MAXIMUM NUMBER OF OUTLIERS IS REACHED
** THE OUTLIER(S) AFTER TIME PERIOD 108 OCCURS WITHIN THE
   LAST FIVE OBSERVATIONS OF THE SERIES. THE IDENTIFIED TYPE
   ANS THE ESTIMATE OF THE OUTLIER(S) MAY NOT BE RELIABLE

```
TOTAL NUMBER OF OBSERVATIONS. . . . . . . . . . . . .         113
EFFECTIVE NUMBER OF OBSERVATIONS. . . . . . . . . . .         112
RESIDUAL STANDARD ERROR (WITHOUT OUTLIER ADJUSTMENT). .  0.790670E+04
RESIDUAL STANDARD ERROR (WITH OUTLIER ADJUSTMENT) . . .  0.431469E+04
--
```

C ADJUSTING THE USER DEFINED MODEL TO OUTLIERS
--
NEW SERIES IN X1, X2, X3.
--
C SAVES THE INFORMATION DURING THE OUTLIER DETECTION PROCESS
--
C X1: NAME USED TO STORE RESIDUALS AFTER OUTLIER ADUSTMENT
--
C X2: NAME FOR THE ADJUSTED SERIES
--
C X3: NAME FOR THE INDICATOR VARIABLES WITH THE TYPE OF THE
--
C OUTLIER.
--
C X2 AND X3 WILL BE NEEDED FOR THE SCAGRAF SESSION TO PRINT
--
C THE OUTLIER PLOT
--
FORECAST MODELSAM.NOFS ARE 4

```
----------------------------------
   4 FORECASTS, BEGINNING AT   113
----------------------------------
 TIME     FORECAST    STD. ERROR    ACTUAL IF KNOWN
  114   69278.7124    4314.6929
  115   69278.7124    4591.2353
  116   69278.7124    4852.0418
```

```
  117  69278.7124   5099.5271
--

C FORECAST OF THE USER DEFINED MODEL
--

OFORECAST MODELSAM.NOFS ARE 4

RESIDUAL STANDARD ERROR (USES DATA UP TO THE FIRST FORECAST ORIGIN)= 4314.7
 TIME    ESTIMATE   T-VALUE    TYPE
   41   24193.603     5.61      IO
   49  -15426.918    -4.07      TC
   63   12000.625     3.16      TC
   71  -15240.739    -3.82      AO
   73  -21780.418    -5.61      TC
   86  -17814.840    -4.13      IO
   87   29235.658     7.49      AO
   94   17504.511     4.44      AO
   97  -14077.231    -3.23      IO
  105   20154.501     6.03      LS
  110   19662.195     4.83      TC
MAXIMUM NUMBER OF OUTLIERS IS REACHED


---------------------------------
 4 FORECASTS, BEGINNING AT  113
---------------------------------

 TIME     FORECAST  STD. ERROR   ACTUAL IF KNOWN
  114   65791.2587  4314.6929
  115   64374.9937  4591.2353
  116   63383.6082  4852.0418
  117   62689.6384  5099.5271
--

C FORECAST FOR OUTLIER ADJUSTED USER DEFINED MODEL
--

RETURN
```

Appendix 4: Data Macro and Output File for Time Series SAM2
The Data Macro (Saved in File SAMHERD2.MAD)

```
==DATA
INPUT SAM2. @
  PREC DOUB. @
  FORMAT FREE(1,132).
 708717
 730216
 881323
```

```
 612434
 823488
 777484
 863486
 920263
 902474
 871949
 910865
 786633
 573128
 624597
 788039
 759398
 688136
 656575
 747635
 672993
 718776
 699143
 667994
1133447
 455398
 479855
 606792
 612942
 449928
 554815
 631958
 504358
 553195
 586408
 633965
 573566
 456985
 312502
 479250
 341501
 359239
 384153
 405545
```

```
346501
389564
420554
389766
384434
231055
252932
333538
342438
303325
298243
311718
281582
387163
345750
344925
341091
303342
319034
338372
350651
400419
End of Data
-- This data macro was created on: 17.05.2001 22:16:22
-- Number of data rows written: 65
Return
```

The Output File for Time Series SAM2 (Saved in File SAM2.OTP)

```
THE SCA STATISTICAL SYSTEM ( RELEASE VI.2B )
 SCA PRODUCT  IDENTIFICATION: GSA, UTS & QPI
 SCA PRODUCT  IDENTIFICATION: EXTENDED-UTS & EXPERT
 SCA SOFTWARE IDENTIFICATION: NOLLAU.H-REFA    (  1041919 )
 SCA SOFTWARE RELEASE DATE:  3/ 1/2001
 SCA SOFTWARE RENEWAL DATE: 12/ 2/2006
 COPYRIGHT (C), SCIENTIFIC COMPUTING ASSOCIATES. ALL RIGHTS RESERVED
 SIZE OF WORKSPACE IS  4000000   SINGLE PRECISION WORDS
 DATE -- 11/26/2001                TIME --  18:38:25
 --
 CALL ALLMACRO. FILE 'SAMHERD2.MAC'
 --
   ASSIGN FILE 12. EXTERNAL 'SAMHERD2.MAD'.
 --
   CALL DATA. FILE 12.
 --
 INPUT SAM2. @
   PREC DOUB. @
```

```
    FORMAT FREE(1,132).

    SAM2    , A    65  BY    1 VARIABLE, IS STORED IN THE WORKSPACE; DOUBLE PRE-
    CISION
    --
    RETURN
    --
    C TIME SERIES VARIABLE SAM2 MAY BE REFERENCED BY POSITIONAL VARIABLE:
    --
    C &V_1.
    --
      CALL ANALYSIS
    --
    C AUTOMATIC MODELING
    --
    &V_1=LN(&V_1)
    --
    IARIMA &V_1


    THE FOLLOWING ANALYSIS IS BASED ON TIME SPAN   1   THRU   65
    THE CRITICAL VALUE FOR SIGNIFICANCE TESTS OF ACF AND ESTIMATES IS 1.731
    SAMPLE ACF OF THE RESIDUALS (** SIGNIFICANT VALUES EXIST **)
    1 - 12     .01 -.14  .00  .03 -.09 -.05 -.02 -.04  .08 -.20  .08  .41
    T-VALUE    .12-1.12  .04  .27 -.72 -.37 -.12 -.32  .64-1.57  .62 3.02

    13 - 24    .14 -.22  .05 -.02 -.11 -.08 -.04 -.11  .03 -.09  .01  .24
    T-VALUE    .88-1.44  .32 -.14 -.66 -.47 -.28 -.68  .17 -.54  .04 1.46

    SUMMARY FOR UNIVARIATE TIME SERIES MODEL -- UTSMODEL
    ----------------------------------------------------------------------
    VARIABLE   TYPE OF    ORIGINAL     DIFFERENCING
               VARIABLE   OR CENTERED
                                            1
      SAM2     RANDOM     ORIGINAL     (1-B  )
    ----------------------------------------------------------------------

    PARAMETER  VARIABLE  NUM./  FACTOR  ORDER  CONS-    VALUE    STD    T
      LABEL     NAME     DENOM.                TRAINT          ERROR  VALUE
      1         SAM2     MA      1       1     NONE    .6336   .0982  6.45

    TOTAL NUMBER OF OBSERVATIONS . . . .          65
    EFFECTIVE NUMBER OF OBSERVATIONS . .          64
    RESIDUAL STANDARD ERROR. . . . . . .  0.186582E+00
    --
    C AUTOMATIC MODELING OF THE LN- TRANSFORMED DATA
    --
    ACF &V_1. MAXLAG IS 36


    NAME OF THE SERIES . . . . . . . . . .        SAM2
    TIME PERIOD ANALYZED . . . . . . . . . 1  TO    65
    MEAN OF THE (DIFFERENCED) SERIES . . .     13.1313
    STANDARD DEVIATION OF THE SERIES . . .      0.3877
    T-VALUE OF MEAN (AGAINST ZERO) . . . .    273.0722


    AUTOCORRELATIONS

      1- 12    .84  .79  .77  .75  .69  .67  .64  .60  .59  .50  .50  .51
      ST.E.    .12  .19  .24  .27  .30  .33  .35  .36  .38  .39  .40  .41
```

```
Q           47.8 90.6  132   173   208   241   271   299   326   345   365   386

13- 24      .43   .33   .32   .25   .18   .15   .11   .07   .05  -.01  -.02  -.04
ST.E.       .42   .43   .43   .44   .44   .44   .44   .44   .44   .44   .44   .44
Q           402   411   420   426   429   431   432   432   433   433   433   433

25- 36     -.11  -.16  -.18  -.22  -.25  -.27  -.29  -.31  -.32  -.34  -.36  -.35
ST.E.       .44   .44   .44   .44   .44   .45   .45   .45   .46   .46   .46   .47
Q           434   437   441   446   454   463   474   487   501   518   537   555


            -1.0 -0.8 -0.6 -0.4 -0.2  0.0  0.2  0.4  0.6  0.8  1.0
            +----+----+----+----+----+----+----+----+----+----+
                                     I
  1   0.84                       +   IXXXXX+XXXXXXXXXXXXXXX
  2   0.79                     +      IXXXXXXX+XXXXXXXXXXX
  3   0.77                  +         IXXXXXXXXXXX+XXXXXXX
  4   0.75                 +          IXXXXXXXXXXXX+XXXXXX
  5   0.69               +            IXXXXXXXXXXXXXX+XX
  6   0.67              +             IXXXXXXXXXXXXX+X
  7   0.64             +              IXXXXXXXXXXXXXXX+
  8   0.60           +                IXXXXXXXXXXXXXX   +
  9   0.59          +                 IXXXXXXXXXXXXXX    +
 10   0.50          +                 IXXXXXXXXXXX       +
 11   0.50         +                  IXXXXXXXXXXX        +
 12   0.51         +                  IXXXXXXXXXXX        +
 13   0.43       +                    IXXXXXXXXXX           +
 14   0.33       +                    IXXXXXXXX             +
 15   0.32       +                    IXXXXXXXX             +
 16   0.25       +                    IXXXXXX               +
 17   0.18       +                    IXXXX                 +
 18   0.15     +                      IXXXX                   +
 19   0.11     +                      IXXX                    +
 20   0.07     +                      IXX                     +
 21   0.05     +                      IX                      +
 22  -0.01     +                      I                       +
 23  -0.02     +                     XI                       +
 24  -0.04     +                     XI                       +
 25  -0.11     +                   XXXI                       +
 26  -0.16     +                  XXXXI                       +
 27  -0.18     +                  XXXXI                       +
 28  -0.22     +                 XXXXXI                       +
 29  -0.25     +                XXXXXXI                       +
 30  -0.27     +               XXXXXXXI                       +
 31  -0.29     +               XXXXXXXI                       +
 32  -0.31     +              XXXXXXXXI                       +
 33  -0.32     +              XXXXXXXXI                       +
 34  -0.34   +                XXXXXXXXXI                        +
 35  -0.36   +                XXXXXXXXXI                        +
 36  -0.35   +                XXXXXXXXXI                        +
--
```

C DEPICTING OF THE ACF FOR THE TIME SERIES LN(SAM1)

--

OUTLIER UTSMODEL. TYPES ARE AO,IO,LS,TC

INITIAL RESIDUAL STANDARD ERROR = 0.18380

```
TIME    ESTIMATE   T-VALUE   TYPE
 24       0.65      4.18     AO
 38      -0.53     -3.60     IO
 49      -0.52     -3.91     IO
```

ADJUSTED RESIDUAL STANDARD ERROR = 0.13283
--

C OUTLIER DETECTION OF THE MODEL CREATED BY THE IARIMA COMMAND

--

OESTIM UTSMODEL. METHOD IS EXACT

THE FOLLOWING ANALYSIS IS BASED ON TIME SPAN   1   THRU   65

SUMMARY FOR UNIVARIATE TIME SERIES MODEL -- UTSMODEL

```
---------------------------------------------------------------------
VARIABLE   TYPE OF     ORIGINAL       DIFFERENCING
           VARIABLE   OR CENTERED
                                     1
 SAM2     RANDOM     ORIGINAL    (1-B )
---------------------------------------------------------------------
```

PARAMETER LABEL	VARIABLE NAME	NUM./ DENOM.	FACTOR	ORDER	CONS- TRAINT	VALUE	STD ERROR	T VALUE
1	SAM2	MA	1	1	NONE	.7011	.0979	7.16

SUMMARY OF OUTLIER DETECTION AND ADJUSTMENT

```
-------------------------------------
TIME    ESTIMATE   T-VALUE   TYPE
-------------------------------------
 13      -0.389     -3.23    IO
 24       0.539      4.66    AO
 25      -0.407     -3.25    IO
 38      -0.547     -4.54    IO
 49      -0.477     -4.61    TC
-------------------------------------
```

```
TOTAL NUMBER OF OBSERVATIONS. . . . . . . . . . . . .       65
EFFECTIVE NUMBER OF OBSERVATIONS. . . . . . . . . . .       64
RESIDUAL STANDARD ERROR (WITHOUT OUTLIER ADJUSTMENT). .  0.188498E+00
RESIDUAL STANDARD ERROR (WITH OUTLIER ADJUSTMENT) . . .  0.120408E+00
```
--

C ESTIMATION OF THE MODEL PARAMETERS INCLUDING THE INFORMATION OF THE
--
C OUTLIERS
--
    NEW SERIES IN V1, V2, V3.
--
C SAVES THE INFORMATION DURING THE OUTLIER DETECTION PROCESS
--
C V1: NAME USED TO STORE RESIDUALS AFTER OUTLIER ADUSTMENT
--
C V2: NAME FOR THE ADJUSTED SERIES
--

C V3: NAME FOR THE INDICATOR VARIABLES WITH THE TYPE OF THE
--
C OUTLIER.
--
C V2 AND V3 WILL BE NEEDED FOR THE SCAGRAF SESSION TO PRINT
--
C THE OUTLIER PLOT
--

FORECAST UTSMODEL.NOFS ARE 4

----
   4 FORECASTS, BEGINNING AT   65

   TIME   FORECAST   STD. ERROR   ACTUAL IF KNOWN

     66   12.7751   0.1204
     67   12.7751   0.1257
     68   12.7751   0.1307
     69   12.7751   0.1356
--
C FORECAST FOR TIME SERIES FOUND BY IARIMA
--

OFORECAST UTSMODEL.NOFS ARE 4

RESIDUAL STANDARD ERROR (USES DATA UP TO THE FIRST FORECAST ORI-
GIN)=0.12041

   TIME   ESTIMATE   T-VALUE   TYPE
     13   -0.389   -3.23   IO
     24   0.539   4.66   AO
     25   -0.407   -3.25   IO
     38   -0.547   -4.54   IO
     49   -0.477   -4.61   TC

----
   4 FORECASTS, BEGINNING AT   65

   TIME   FORECAST   STD. ERROR   ACTUAL IF KNOWN

     66   12.7821   0.1204
     67   12.7825   0.1257
     68   12.7827   0.1307
     69   12.7829   0.1356
--
C FORECAST OF THE LN- TRANSFORMED TIME SERIES SAM2 INCLUDING THE OUTLIER
C INFORMATION
--

   RETURN
--
   STOP

THE CURRENT SCA SESSION IS TERMINATED.
THE SIZE OF THE WORKSPACE USED IS 15216 WORDS.

Appendix 5: Forecasting Irregular Time Series

As introduced in chapter 2 irregular time series have the proberty that $y_t = 0$ appears very often and also repeated in the data. Especially spare parts show this behaviour. Conventional forecasting methods, that work well under normal conditions, give inaccurate results with irregular data. Some organizations used self- made techniques but those methods couldn't co-operate with large numbers of items.

The NSF study, conducted by a team from Rensselaer Polytechnic Institute and Smart Soft-ware Inc. examined 28 000 commercial data series-inventory items from nine companies in the U.S.A. and Europe, representing the aircraft, high-tech electronic components, marine equipment and other capital equipment industries. Key findings from this research are as fol-lows:

1. Traditional forecasting methods fail with intermittent data because they assume that the probability distribution of demand over a lead-time (lead-time demand) will resemble a "nor-mal" bell-shaped curve. It doesn't, especially in the case of service parts, whose lead-time demand can exhibit odd shapes.

2. Traditional computerized forecasting methods fail because they try to identify patterns in the data, such as seasonality and trend. Intermittent demand data exhibit no such easily identi-fiable patterns.

3. Both exponential smoothing and a variant of exponential smoothing, developed by the statistician J.D. CROSTON in 1972, are effective in forecasting mean (average) demand per period when demand is intermittent. But, neither CROSTON's method nor exponential smoothing accurately forecasts the entire distribution of demand values over a lead time, which is especially necessary when estimating customer service-level inventory requirements (e. g., a 95 or 99 percent likelihood of not stocking out of an item) for satisfying total demand over a lead time. Lead-time demand forecasts are key inputs to inventory models, which rec-ommend optimal procedures for inventory management, such as the size and timing of re-plenishment orders, i.e. order quantities and reorder points.

4. A new method that incorporates *bootstrapping* provides fast, realistic forecasts of intermittent product demand. Bootstrapping is a statistical method that forecasts both average demand per period and customer service-level inventory requirements by using samples of demand history to create thousands of realistic scenarios that show the evolution of cumulative demand over a lead time.

You will find more detailed information on the WebPage [wwwsm].

References

[Ch82]   Chatfield, Christopher: Analyse von Zeitreihen. Carl Hauser Verlag München Wien, 1982

[Ba00]   Baier, Gerhard: Knowledge Engineering auf der Basis integrierter, betriebswirtschaftlicher Standardsoftware. Diplomarbeit im Fachbereich BWL, Schwerpunkt Logistik der FH-Regensburg, 2000, Betreuer Prof. Dr. H.-G. Nollau

[NoS92]   Nollau, H.-G.; Schmager, B.: Neue Entscheidungsinstrumente für das Produktionsmanagement. In: REFA-Nachrichten, Nr. 03/1992, S. 27–35

[NoH01]   Nollau, H.-G.; Hauser, Thomas: Geschäftsprozeßoptimierung in der Beschaffungslogistik eines mittelständischen Unternehmens. Schriftenreihe zur KGVL®, Bd. 2, RKW Verlag, 2001

[No83]   Nollau, H.-G.: Beiträge zur Zeitreihenanalyse von Belastung und Beanspruchung in der Ergonomie. Dissertation, Darmstadt 1983

[N00]   Nollau, H.-G.: Grundzüge der KGVL®. Schriftenreihe zur KGVL® Bd.1. RKW Verlag 2000

[Hi00]   Hirn, Josef: Optimierung der logistischen Prozesse bei einem mittelständischen Unternehmen des Maschinenbaus. Diplomarbeit im Fachbereich BWL, Schwerpunkt Logistik der FH-Regensburg, 2000, Betreuer: Prof. Dr. H.-G. Nollau

[Hu91]   Hudak, Gregory B.: The SCA Statistical System, SCA Corp. 1991

[SCA01]   SCA Corp.: SCA WorkBench 2000 (4.2) User's Guide, SCA Corp., 2001

[SCA95] SCA Corp: SCA GRAPHICS Package User's Guide Version 2, SCA Corp.,
         1995

[LML01] Lon-Mu Liu: Data mining of Time Series: An Illustration Using Fast- Food
         Restaurant Franchise Data. Department of Information and Decision Sciences,
         University of Illinois at Chicago, 2001

[LML00] Lon-Mu Liu: Forecasting and Time Series Analysis using the SCA Statistical
         System, Volume 1, SCA Corp., 2000

[L00]   Lon-Mu Liu: Forecasting and Time Series Analysis using the SCA Statistical
         System, Volume 2, SCA Corp., 2000

[LML05] Lon-Mu Liu: Time Series Analysis and Forecasting. Scientific Computing
         Associates Corp., 2005

[G00]   Götze, Wolfgang: Techniken des Business- Forecasting. R. Oldenbourg Verlag
         München Wien, 2000

[SPP98] Silver, E. A.; Pyke, D. F.; Peterson, R.: Inventory Management and Production
         Planning and Scheduling. John Wiley, New York 1998, Third Edition, Kap. 4
         Forecasting

[SS97]   Schlittgen, Streitberg: Zeitreihenanalyse. Oldenbourg Verlag, München Wien,
         1997, Kap. 6.3

References of the Internet

[wwwsc]   www.scausa.com

Website of Scientific Computing Associates Corp., the producer and seller of
the SCA-Statistical System.

[wwwsm]   www.smartcorp.com

Website of Smart Software Inc. with information about this US-company and
an interesting article by C. N. Smart called Smart White Paper and entitled *Ac-*
curate Intermittent Forecasting for Inventory Planning , where the Smart-Wil-
lemain Method, based on bootstrapping, is shortly described and illustrated
with an example for computing a lead time demand distribution.

ECONOMY AND LABOUR

Herausgegeben von EUR ING Prof. Dr.-Ing. Hans-Georg Nollau
FBCS, Roßdorf

Band 10
Andreas Tiemann
Internetbasiertes Projektmanagement am Beispiel des Bauwesens
Lohmar – Köln 2006 ◆ 212 S. ◆ € 45,- (D)
ISBN-13: 978-3-89936-504-7 ◆ ISBN-10: 3-89936-504-6

Band 11
Amelie Heerd
Beschaffungslogistik in Beispielen
Lohmar – Köln 2006 ◆ 134 S. ◆ € 38,- (D)
ISBN-13: 978-3-89936-532-1 ◆ ISBN-10: 3-89936-532-0

Band 12
Hans-Georg Nollau, Matthias Neumeier und Emilio Sabatino
Dokumentenmanagement in der Logistik
Lohmar – Köln 2006 ◆ 200 S. ◆ € 44,- (D)
ISBN-13: 978-3-89936-533-7 ◆ ISBN-10: 3-89936-533-X

Band 13
Hans-Georg Nollau, Stefan März und Andreas Wolf
Prozesssimulation mit der Software DOSIMIS
Lohmar – Köln 2008 ◆ 210 S. ◆ € 55,- (D) ◆ ISBN 978-3-89936-717-1

Band 14
Hans-Georg Nollau, Uli Gottfried
Entscheidungskompetenz durch Anwendung der Vektor-Nutzwertanalyse
Lohmar – Köln 2009 ◆ 268 S. ◆ € 57,- (D) ◆ ISBN 978-3-89936-761-4

Band 15
Hans-Georg Nollau und Matthias Neumeier
Logistikfallstudien und Risikomanagement
Lohmar – Köln 2010 ◆ 248 S. ◆ € 56,- (D) ◆ ISBN 978-3-89936-903-8

Band 16
Hans-Georg Nollau and Carmen Zech
Forecasting: A Challenge for True Statisticians
Lohmar – Köln 2010 ◆ 168 S. ◆ € 47,- (D) ◆ ISBN 978-3-8441-0009-9

JOSEF EUL VERLAG